T0374936

Learning to Imagine

Learning to Imagine

The Science of Discovering New Possibilities

Andrew Shtulman

Harvard University Press
CAMBRIDGE, MASSACHUSETTS LONDON, ENGLAND 2023

Printed in the United States of America
First printing

Library of Congress Cataloging-in-Publication Data

Names: Shtulman, Andrew, author.
Title: Learning to imagine : the science of discovering new possibilities / Andrew Shtulman.
Description: Cambridge, Massachusetts ; London, England : Harvard University Press, 2023. | Includes bibliographical references and index.
Identifiers: LCCN 2023002236 | ISBN 9780674248175 (cloth)
Subjects: LCSH: Imagination. | Cognitive learning. | Learning, Psychology of.
Classification: LCC BF408 .S4517 2023 | DDC 153.3/2—dc23/eng/20230415
LC record available at https://lccn.loc.gov/2023002236

To Carina
I can't imagine a better sister

Contents

Learning to Imagine

1 Our Unimaginative Imagination

In *The Little Prince,* the narrator recounts a time in his youth when he drew a picture of a boa constrictor that had eaten an elephant. He proudly showed it to the grownups around him and asked if they were frightened, but they didn't understand why they should be frightened of a hat. Frustrated by the grownups' lack of imagination, he drew a picture of the boa as if it had been x-rayed so the elephant could be seen more plainly, but the grownups' response was that he should put his drawings aside and pursue more useful skills like arithmetic and grammar.

When the narrator grew up, he kept his first drawing and used it as a litmus test to see who might view it properly, as an elephant-engorged boa. But his fellow grownups always described it as a hat, and when they did, he resolved never talk to them about snakes or forests or stars. "I would bring myself down to his level," explains the narrator. "I would talk to him about bridge, and golf, and politics, and neckties. And the grownup would be greatly pleased to have met such a sensible man."[1]

The idea that children are imaginative but adults are not is a common theme in popular fiction. In *Peter Pan,* only children can go to Neverland and meet the fairies, pirates, and mermaids who live there. Grownups cannot "because they are no longer gay and innocent and heartless. It is only the gay and innocent and heartless who can fly."[2] In *Charlie and the Chocolate Factory,* only children are able to wrap their minds around Willy Wonka's amazing inventions. Wonka is intent on leaving his factory to a child, rather than a grownup, because "a grown-up won't listen . . . he won't learn."[3] In *Mary Poppins,* only children are able to hear what the trees and the birds say, and only children are able to understand the language of the sun and the stars. "Do you really mean we won't be able to hear that when we're older?" asks one of the children. "You'll hear all right," explains Mary Poppins, "but you won't understand."[4]

In these stories, children are open to possibilities that elude the adult mind. They grasp extraordinary ideas that adults cannot fathom and have clever insights that adults do not appreciate. It's a theme that resonates with everyday observation of children's imaginative activities. Children play elaborate make-believe games with their toys and role-playing games with their peers. They build forts, assemble costumes, bake mud pies, and construct block towers. They create imaginary friends and chart imaginary worlds. They believe in fantastical beings, like Santa Claus and the Tooth Fairy, and they are convinced that magic is real. They are ardently curious about the way things work and keenly inventive in explaining why.

On closer inspection, though, many of these activities are more mundane than they first appear. Children spend most of their pretend play doing realistic things, like cooking and cleaning, and they eagerly abandon pretend play if allowed to do those things for real.[5] Their homespun creations—forts, costumes, and the like—are usually replicas of adult artifacts, created to imitate the adult world rather than depart from it.

In fact, children prefer imitation to innovation in general. When playing games, they stick closely to the rules and are offended by anyone who might attempt to change them.[6] When drawing pictures, they focus on ordinary objects and are baffled by the request to draw something that doesn't exist.[7] Some children do invent imaginary friends and imaginary worlds, but they are imaginary in the sense that they don't exist, not that they couldn't. Imaginary friends behave a lot like real friends, and imaginary worlds are as rule-bound as the real world.[8]

Although children do believe in fantasy characters like Santa Claus and the Tooth Fairy, these beliefs are not spontaneous inventions; society colludes to convince children they are real.[9] In the absence of such collusion, children are highly skeptical of extraordinary ideas. They claim that events that violate their expectations cannot happen in the real world, whether those events are magical or merely improbable.[10] They deny, for instance, that a person could own a lion for a pet or wear shoes in the bathtub. Prompting children to imagine these events in their mind does not increase their willingness to accept them as possible, nor does providing children with an explanation for how the events might occur.[11] When children say that magic is real, it's not because they think nothing is truly impossible; rather, it's because many of the events they thought were impossible turn out to be real.[12]

And yes, children are curious, but they are most curious about things they largely understand. They explore the world in ways intended to confirm their expectations, not challenge them. When exploring a new object or situation, they repeat what they see others do, failing to discover anything they did not expect to discover.[13] If they do discover something unexpected, such as a surprising feature of a tool or a surprising consequence of an action, they often ignore the discovery or write it off as unimportant.[14] In the rare instance that children attend to an unexpected discovery, they may explain it in a creative fashion, but creative in the sense of unusual rather than insightful.[15] Children rarely intuit novel causal principles on their own.[16]

Forty years of research on how people reason about novel possibilities reveals that the glorification of children's imagination is misguided. Children are no more imaginative than adults. Quite often, they are less imaginative. Children have the capacity to contemplate hypothetical ideas and counterfactual events, but they do not have the knowledge or expertise to use that capacity as effectively as adults.

The problem is that our imaginations are firmly anchored in the status quo. People of all ages, adults included, have difficulty contemplating alternatives to reality that differ by more than a few details. If, for instance, we are asked to draw an imaginary animal, we usually draw a real animal and then add an extra feature, like an extra eye or an extra limb.[17] If asked to finish a fictional story, we default to commonplace endings and ignore more creative options.[18]

We struggle to apply familiar tools to novel problems, fixating on what we know the tools are for, and we struggle to apply novel tools to familiar problems, overlooking their relevant affordances.[19] We rarely see how a problem-solving strategy illustrated in one context could be of use in any other.[20] We stick with hypotheses for why something happened even after they've been refuted, and we have difficulty entertaining more than one hypothesis at a time.[21] New technologies are dismissed as unsafe, and new scientific discoveries are dismissed as unreliable.[22] We fail to discern better ways of making a calculation if we have a method that suffices.[23] And we fail to appreciate ethical or legal principles that challenge our understanding of everyday morality.[24]

There is room for innovation in everything we do—cooking, cleaning, writing, drawing, navigating, negotiating—but we rarely innovate. We stick with what we know, making small adjustments to fit the situation but no revolutionary changes. Such changes require sustained effort and reflection. We have to acquire the right knowledge and cultivate the right habits of mind. Imagination, like any other faculty, can be developed and refined, and this book tells the story of how.

Imagination's Purpose

What is imagination? What is it for? Our stereotype of imagination is well captured by a 2020 commercial for Adobe software. The

Figure 1.1 Imagination allows us to contemplate fantastical possibilities, like children riding on clouds, but it did not evolve for this purpose. (Courtesy of the author)

commercial showcases Adobe's image-editing features by cycling through a series of eye-popping pictures, including a mermaid hugging a teddy bear, a child riding a cloud, a whale swimming over a sunken city, and an astronaut discovering a flower on a distant planet. The pictures blend into one another with splashes of color and shifts of perspective, keeping beat with the song "Pure Imagination" from the movie *Charlie and the Chocolate Factory* (1971). "Come with me, and you'll be in a world of pure imagination," goes the song. "We'll begin with a spin, traveling in the world of my creation. What we'll see will defy explanation."

This commercial implies that imagination is for creating rule-bending, mind-blowing ideas. But is that what imagination is really for? Did imagination evolve so we could paint pictures of mermaids hugging teddy bears and write stories about children riding clouds?

Certainly not. Imagination evolved for planning and predicting, not innovating.[25] Every time we entertain a thought that transcends what we are currently perceiving, we are using imagination. Thinking of mermaids requires imagination but so does thinking of past vacations, distant friends, or future meetings. Almost all mental life requires traveling beyond the here and now to contemplate what was, what will be, what might be, what should be, and what could have been. Planning your day requires imagination, as does planning your route, retracing your steps, deciding what to wear, rehearsing what to say, anticipating a response, estimating a value, remembering a joke, telling a story, telling a lie, or daydreaming.

Life is a series of problems—what to eat? where to go? who to ask?—and solving those problems requires entertaining multiple possibilities and then selecting the best option among them. The possibilities we entertain are inherently constrained by those we have already encountered. What we imagine ourselves doing in the future is usually some permutation of what we have done in the past.

The connection between remembering the past and forecasting the future runs deep, both developmentally and neurologically.[26] Children who accurately remember what they did yesterday are more accurate at predicting what they will do tomorrow.[27] Children who accurately recall how they performed a task in the past are better at predicting how they will perform that same task in the future under different circumstances.[28] Thinking about the future recruits regions of the brain that are also involved in thinking about the past, such as the hippocampus.[29] If those regions are damaged by an accident or a stroke, the result is not only amnesia for past events but also difficulty contemplating future events.[30]

Contemplating future events is a prerequisite for intelligent, flexible behavior. It is a hallmark of human thought—but is not exclusive to humans. It also characterizes the thought of other intelligent animals, such as scrub jays.[31] Scrub jays are part of the corvid family, which includes crows, ravens, and rooks. When foraging, they hide excess food to retrieve at a later date. Scrub jays have excellent memories not only for where they have hidden their food but also for what they hid and when. They are particularly fond of worms, but these delicacies perish fast, so when scrub jays hide both worms and nuts, they return for the worms first. And if the first worm they retrieve has decayed, then they assume all the worms they've hidden have decayed and switch to retrieving nuts instead.

Imagination thus affords mental time travel. It allows us to move forward in time or backward in time or jump to another time line altogether. We can contemplate events that have happened, will happen, or did not happen but might have if the circumstances had been different. The latter are known as counterfactuals: the events that underlie our regret for lost opportunities, our relief at avoiding misfortune, and our surprise that events turned out one way rather than another. We contemplate counterfactuals not just because we enjoy exploring alternative time lines but because they help us learn more about our current time line so we might steer ourselves toward future successes and away from future failures.[32]

Take, for instance, the goal of cooking a successful meal. When we discover that a meal we've cooked is not a success, we could throw up our hands and hope for the best next time, or we could do something more proactive: we could retrace our steps in the kitchen and attempt to pinpoint what went wrong. Were the ingredients fresh? Were they combined in the right order? Were they cooked for the right amount of time? We engage in this mental simulation effortlessly and spontaneously, but it requires some sophisticated accounting. We have to create several distinct models of reality—a model of what happened, a model of what we wanted to happen, a model of what we could have done differently, and a model of what might have happened instead—and then coordinate those models to isolate the events we want to change in the future, the causal events.

There are other ways to identify causal events, such as cooking the same meal over and over again, with slight permutations, or consulting others who have cooked the same meal and then comparing their process to ours. These methods might prove useful, but they might also waste our time. Without speculating on the events that caused our cooking disaster, we could spend months comparing one meal preparation to another before hitting upon the critical difference between a successful preparation and an unsuccessful one. Counterfactuals let us tinker with reality without the hassle of producing or investigating real outcomes. Our counterfactual simulations may not always be accurate, but they help us narrow a vast space of possibilities within minutes rather than months.

Using counterfactuals to improve causal reasoning does not have to be instructed. This skill emerges on its own, early in development. Children as young as three recognize that a person with muddy shoes would not have muddied the floor if they had left their shoes outside.[33] Slightly older children can discriminate between counterfactuals that would change the course of events from those that would not, acknowledging, for instance, that a person would not be cold if they had worn a jacket but would still be cold if they had worn a shirt of a different color.[34]

A child's ability to discriminate between effective and ineffective counterfactuals develops alongside their ability to explain why events happen. The better they can imagine how things might have been different, the better they can identify the causes of things that actually happened.[35] Children who generate appropriate counterfactuals are also better at generating future hypotheticals, recognizing not just how past events could be altered but also how future events could be brought about, given the right circumstances.[36]

Space, time, and causation are all manipulable within the workspace of imagination. But the utility of such manipulations has its limits. Counterfactuals that alter too many facts will no longer be informative. Goals that depart too far from current circumstances will no longer be achievable. And predictions that stray too far from prior outcomes will no longer be correct. Imagination may have evolved for contemplating alternatives to reality, but we use it most naturally to contemplate close alternatives, like preparing a different meal, rather than far alternatives, like riding on clouds. In fact, the alternatives we most commonly consider hew so close to reality that we do not think of them as products of imagination. When we use imagination to contemplate far alternatives—to innovate or fabricate—we're not tapping into an innate appreciation of the extraordinary; we're coopting a tool designed to explore the ordinary.

Imagination's Structure

Imagination is limited in scope because it is limited in structure. When contemplating alternatives to reality, we manipulate some dimensions but fail to consider many others. We fixate on possibilities that are physically plausible, statistically probable, socially conventional, and morally permissible. When told about possibilities that violate such regularities, we usually balk at their suggestion, denying they could happen. Our ideas about what could happen are firmly rooted in what we expect to happen.[37]

Imagine, for instance, you're at a restaurant and the diner next to you discovers a fly in his soup. What might he do next? You would expect him to ask the waiter for a new bowl of soup or possibly leave the restaurant. You wouldn't expect him to eat the soup with the fly in it or pull a new, uncontaminated bowl from under the table. You also wouldn't expect him to swap his bowl for his date's when she is in the bathroom or recite an incantation over the bowl to remove the fly by magic. What's more, the events you expect to happen are the only events you entertain. You don't imagine the diner pulling a fresh bowl of soup from under the table and then dismiss the possibility as unlikely—you don't entertain that possibility at all.

Of course, you shouldn't entertain such possibilities, at least not when navigating real-world situations. They would only interfere with proper decision making. The eighteenth-century philosopher Count Buffon called highly improbable outcomes "morally impossible" in the sense that we are obliged not to consider them.[38] But this mindset, when applied broadly, conflates possibility with probability. We only consider what's probable as possible, either when generating ideas or when considering the ideas of others.

This mindset is most apparent in young children, who are quick to dismiss the unexpected as impossible. For example, in one study four-year-olds were told about commonplace problems and asked to contemplate various solutions to those problems, some more unusual than others.[39] One of the problems was about a girl named Melissa who didn't like to go to school because she missed her mother too much. What could Melissa do to solve her problem? Could she and her mother agree to do something special after school to take her mind off her worries? Could she wear her pajamas to school for comfort? Could she bring her mother to school to attend classes with her? Could she lie to her mother and tell her that school is closed today so she doesn't have to go? Could she snap her fingers and make it Saturday so school is actually closed?

Four-year-olds thought only the first solution (the afterschool treat) could happen in real life; the rest were judged impossible. Children claimed not only that these events could not occur in real life but also that it would take magic to make them happen.

Changing the day of the week is of course impossible, but the other solutions are not. There are reasons why a student might not want to wear pajamas to school or bring her mother to school or lie to her mother about school being closed, but these reasons do not preclude the events from occurring. Children's earliest intuitions about possibility conflate what could happen with what should happen. Children are not confused about possibility itself; they recognize that multiple outcomes are possible in most situations and that low-probability outcomes are different than zero-probability outcomes.[40] Still, they tend to mistake reasons why events *do not* occur for reasons why they *could not* occur.

Adults recognize that improbable, unconventional, and immoral events are possible, but we too conflate these distinctions when making snap judgments.[41] If given only a couple of seconds to decide whether an action is possible, adults judge immoral actions and unconventional actions as impossible, just like children. Our judgments of possibility converge with our judgments of probability and permissibility the less time we have to reflect. We claim not just that lying to your mother is wrong but that a person couldn't lie to their mother, despite many observations (and personal experiences) to the contrary.

Wouldn't, shouldn't, and couldn't are distinctions that require learning and reflection; they are not inherent features of imagination. As further illustration, consider the following riddle: "A young woman attends her mother's funeral. At the funeral, she meets a man she doesn't know. She discovers that he is amazing and falls in love with him immediately. But she forgot to ask for his name or number and afterward could not find anyone who knew who he was. A few days later, the woman kills her own sister. Why did she do it?"

This riddle has been circulated on the internet as a test for psychopathy, as only psychopaths would infer that the woman killed her sister in the hope of meeting with her love interest at another family funeral. The reasoning is that if he appeared at her mother's funeral then he might appear at her sister's funeral as well. This test is apocryphal—it doesn't actually diagnose psychopathy—but its perceived credibility highlights how notions of possibility are constrained by notions of permissibility. The average person does not entertain the possibility of murdering a sibling for instrumental reasons and then reject it on moral grounds; we never entertain such a possibility at all.

It may be good that imagination steers clear of immorality, but it's not so good that it steers clear of unconventionality and improbability as well. The improbable event of traveling faster than a horse was once considered impossible, as was traveling by air or traveling into space.[42] Before the advent of trains, planes, and rocket ships, there were good reasons to think that people could travel only so far and only so fast. But these reasons were empirical, not logical. They could be altered, and they were. Imagination, on its own, lumps the improbable with the impossible, but we can coordinate imagination with other faculties—namely, knowledge and reflection—to disentangle the two. Unstructured imagination succumbs to expectation, but imagination structured by knowledge and reflection allows for innovation.

Expanding Imagination

The focus of this book is on how we discover new possibilities, be it new technologies, new scientific principles, or new ethical commitments. Imagination is necessary for these discoveries but not sufficient; knowledge is required as well. Einstein famously quipped that "imagination is more important than knowledge," but the two are not distinct. Quite the opposite, they are intrinsically linked.[43] Knowledge supports imagination by providing the raw materials for generating alternative possibilities. You have to represent reality before you can tinker with it, to know the facts before you can entertain counterfactuals. But knowledge also impedes imagination by limiting the scope of our investigation, leading us to false starts or down blind alleys.

As an illustration, try your hand at the Remote Associates Test, which is commonly used to measure creativity. In this test, you're given three words and must find a fourth word that can be combined with each to form a compound word or phrase. What word, for instance, can be combined with "house," "guard," and "hot"? "Safe" can be combined with "house" to make "safe house" and with "guard" to make "safeguard," but it can't be combined with "hot;" "safe hot" and "hot safe" are not valid compounds. What word can be combined with all three?

This task, which is hard, is made even harder by considering a nonstarter solution like "safe." We fixate on "safe" and are unable to consider alternatives.[44] To solve this task, we need to clear our mind of "safe" but not clear our mind completely. The solution won't surface on its own. Rather, we have to identify options that work for one word and then apply that option to the others. Focusing on "house," we can come up with several candidates: "tree" as in treehouse, "boat" as in boathouse, or "white" as in White House. But these candidates don't work when tested against the other words: "tree" doesn't pair with hot or guard; "boat" pairs with guard but not hot; and "white" pairs with hot but not guard.

The solution—"dog," as in "doghouse," "guard dog," and "hotdog"—is discovered by actively consulting our knowledge of compound words. We cannot solve this problem if we do not know enough compound words or if we do not access them in a useful manner. The same is true for any other task that requires coordinating multiple possibilities. We must consult some knowledge base—animal facts, historical events, product designs, cooking procedures, painting styles, mathematical operations—and we must consult it purposefully and productively. What we know and how we know it provide the foundation for what we can imagine.

In this way, knowledge acts a double-edged sword. It hampers imagination when it leads us to search for solutions where none can be found, but it facilitates imagination by allowing us to change course and search somewhere else. The psychologists Douglas Hofstadter and Emmanuel Sander have likened the relation between imagination and knowledge to that of a train and a track.[45] The track constrains where the train might go but is necessary for the train to move at all. If we wish to move the train in a new direction, then we should lay down new tracks, not destroy the tracks that already exist. Trains can switch tracks, after all.

Building on that metaphor, we can think of knowledge as a familiar path through a landscape of ideas. This path guides imagination from one idea to the next but does not wholly constrain it; imagination can be used to extend the path itself, breaking new ground by contemplating new possibilities. This book is organized in terms of how we break that ground—how we expand the scope of imagination by contemplating three types of possibilities: examples, principles, and models.

Expanding imagination by example is learning about a possibility precluded by our own imagination but realized by the imaginations of others, including new possibilities conveyed through testimony (chapter 2), technology (chapter 3), and empirical discovery (chapter 4). Expanding imagination by principle is generating a new collection of possibilities from an abstract schema, including such schemas as scientific principles (chapter 5), mathematical principles (chapter 6), and ethical principles (chapter 7). Expanding imagination by model is learning about familiar possibilities by immersing ourselves in a world of alternative possibilities, including pretense (chapter 8), fiction (chapter 9), and religion (chapter 10).

Examples are the simplest way to expand imagination. Rather than toil to discover new ideas, we take advantage of ideas already discovered by someone else. When we hear of an event that others have witnessed, interact with a technology that others have invented, or learn of a discovery that others have made, we are immediately transported to a new location in the landscape of ideas, often far from the path we've been traveling. The challenge for imagination is returning to that familiar path—forging a connection between what we have learned and what we already know.

Some people are more successful at this task than others. The more we've explored a landscape, the easier it is to add a new possibility to it. Conversely, the less we've explored a landscape, the easier it is to give up and reject the new possibility outright. The chapters in the first section of this book will outline strategies people use to connect new possibilities to prior knowledge, beginning with children's strategy of simply denying the possibility of anything unexpected.

Learning new principles is a more powerful way of expanding imagination, as principles are more abstract and more generative than solitary examples. Learning that dolphins are mammals may challenge our understanding of marine life, but learning the principle of common descent challenges our understanding of all living things, dolphins included. Learning that snow boots are more expensive in winter may shed light on the pricing of seasonal apparel, but learning the principle of supply and demand sheds light on the pricing of all products, from snow boots to snow cones.

Examples pluck us from a familiar path within the landscape of ideas and transport us to a new location, but principles provide us with the tools for extending the path ourselves. It's the difference between being air-dropped into a jungle and being given a machete to hack our way through. Principles yield many cognitive benefits, but they also impose costs. Principles can take months to learn and years to apply, and their application can block the application of other, more appropriate principles. Everything looks like a nail to those with a hammer. On balance, though, principled knowledge is far superior to unprincipled knowledge in generating new ideas.

Models, like principles, are another means of expanding imagination more powerful than examples. A model is a simplified version of reality that can be manipulated in ways that reality cannot. We learn from models by intervening on them—tweaking a condition, adding a parameter, or suspending an assumption—and then watching how their operations unfold. Models are substantively different from examples and principles because they allow us to simulate reality, not just represent it. Simulations reveal the unknown consequences of our beliefs and suggest new means for achieving our goals.

Models abound in all areas of thought, including those covered in the chapters on math, science, and technology, but the chapters in the final section of this book will focus on models of an experiential nature: pretend play, fictional narratives, and religious cosmologies. Sometimes we create our own models, and sometimes we inherit models from others. Either way, learning from a model is not child's play. It takes skill and effort to learn more from a model than what we already know.

In covering different mechanisms for expanding imagination, this book will also cover different forms of imagination, including historical imagination, technological imagination, mathematical imagination, and moral imagination. Each form has its own conceptual foundations and its own developmental trajectory. Mathematical imagination is grounded in the logic of space and quantity and develops through the discovery of new numbers and operations. Moral imagination, on the other hand, is grounded in interpersonal relations and develops through the discovery of new obligations and permissions. Differences in content from one domain to another lead to differences in how we contemplate new possibilities, from the information we recruit, to the obstacles we face, to the support we require.

Our focus, throughout the book, will be on ideas rather than events. Contemplating how life might be different if we had pursued more education is an act of imagination, but it is personal and subjective, grounded in the details of one's own experience. This book will focus on possibilities of a more objective nature, grounded in collective knowledge and relevant to anyone.

In this same vein, we will focus on knowledge rather action. Mentally simulating a dance, song, or sport requires imagination but of a kind derived from memory and motor planning rather than reasoning and reflection.

Finally, our attention will be devoted to the psychological processes underlying imagination, not historical or sociological processes. We will focus on how individuals look beyond the familiar and the expected to contemplate the unfamiliar and the unexpected. This endeavor could yield insights that no one has ever contemplated, anywhere at any time, but more typically it yields insights already appreciated by humanity at large. Culturally groundbreaking ideas lay the foundation for cognitively groundbreaking ones, as when we learn such cultural innovations as addition and subtraction, integers and fractions, inertia and momentum, equilibrium and feedback, infection and inoculation, symbiosis and parasitism, correlation and causation, currency and trade, equity and equality, or democracy and justice. Humankind may know these ideas, but individual humans must rediscover them, one idea at a time.

This tension between individual and cultural innovation complicates the relationship between imagination and its close ally creativity. If we identify creativity with contributing to the collective knowledge of humankind, then humans are rarely creative. Hardly anyone advances human knowledge, and the advances we do make are usually so slight as to go unnoticed. Children, in particular, would never be deemed creative. As the psychologist Michele Root-Bernstein points out, "Children may reinvent imagined wheels new to themselves, but rarely do they invent effective technology or explore the full value of an artistic form. A child is almost never culturally creative."[46]

If, on the other hand, we identify creativity with discovering ideas that transcend the limits of personal experience, then humans are regularly creative, children included. The key is to supplement imagination with examples, principles, and models—cognitive tools that help us forge a path from familiar facts to counterfactuals, from everyday observations to alternate realities.

A Work in Progress

Discovering new possibilities can be exciting, even joyful. It's what makes many videogames fun to play. Consider the Nintendo classic *The Legend of Zelda*. This game takes place in a rectangular world measuring 256 spaces in length and 88 spaces in width. Some spaces comprise the path used to get around the world. Other spaces comprise the terrain, including rocks, trees, and water. Hidden in the terrain are a variety of secret objects, some of which are essential for winning the game, but these objects cannot be accessed without special tools. You need bombs to explode the rocks, a candle to burn the trees, and a raft to cross the water.

Each tool opens a range of possibilities that the other tools do not, though most possibilities bear no fruit. There are hundreds of rocks to bomb and bushes to burn but only a handful of treasures hidden beneath. Still, players persist in their bombing and burning because each discovery facilitates further discoveries: a new weapon, a new passage, a new level.

Quest games like *The Legend of Zelda* embody many of the principles that govern the discovery of new possibilities. Players are initially restricted to a path of known possibilities within a larger landscape of unknown or inaccessible ones. The quest requires that players expand that path, and they do so with tools. Some tools, like warp zones and secret passages, transport players to new areas of the game world, similar to how examples transport us to new locations in a landscape of ideas. Other tools, like bombs and rafts, allow players to carve new paths for themselves, similar to how principles allow us to carve paths into new conceptual terrains. These tools have limited applicability, though. You

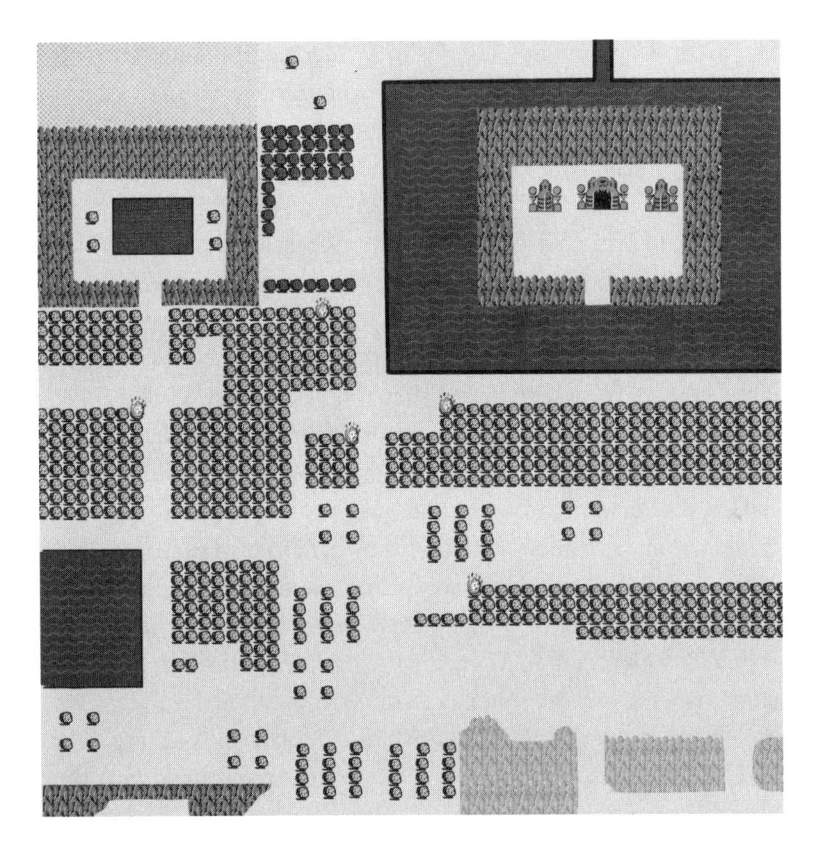

Figure 1.2 The discovery of new ideas resembles the discovery of new resources in a quest game. Our search is initially restricted to a familiar path, but that path can be expanded with specialized tools that uncover possibilities hidden from view or blocked from access. (Courtesy of the author)

can bomb your way through a mountain, but you can't raft over it; and you can raft to an island, but you can't bomb your way there.

A third class of tools, like maps and charts, provide a representation of the landscape that allows players to navigate more effectively or discern possibilities that were unobservable from the ground. These tools correspond to models, which afford not only a new perspective but also the power and efficiency of vicarious exploration.

Discovering new possibilities in a videogame is fun, but it also takes work. Unlike a film or book, where the possibilities unfold before our eyes, we must actively discover the possibilities for ourselves, acquiring the right tools and applying them in the right way. Discovering new possibilities in the real world is similarly labor intensive. Appreciating a groundbreaking idea—example, principle, or model—takes time and effort, as does connecting that idea to what we already know. These activities do not strike us as "imaginative" because we think of imagination as a free resource, a trait we are born with rather than a skill we must develop.

Conceptualizing imagination as a skill, not a trait, requires abandoning several myths about its operation. One myth is that imagination thrives on ignorance: the less you know, the farther your imagination can wander, untethered by hard facts or rigid principles. This myth has permeated many schools and is responsible for the increasingly popular practice of eschewing instruction in favor of self-guided exploration. Many instructors believe it's best not to tell students what to learn so they can discover the ideas for themselves. Yet research on this type of learning—"discovery learning"—has yielded poor results.[47] Students discover little more than what they already knew or what they falsely believed, as we'll discuss in later chapters. Prior knowledge might constrain the search for novel possibilities, but the search itself requires knowing where to begin.

A related myth is that education is the enemy of imagination. If children begin life as great innovators but end up as narrow-minded bean counters, then education must be the culprit. Schools are criticized for making children color within the lines and think inside the box, thus squelching their inherent creativity.

There is some truth to the idea that education routinizes behavior. Children who are taught how to play with a toy tend to repeat the actions they've been taught and fail to discover new affordances.[48] Children who are shown how to solve an unfamiliar problem, like how to retrieve a prize from a locked box, repeat the solution verbatim, even when it contains extraneous steps they would never make on their own.[49] And when children are presented with problems that require using familiar tools in novel ways, such as using a box as a platform rather than as a container, younger children solve these problems faster than older children, presumably because older children fixate on what they know the tool is for.[50]

But do these observations imply that children would be better off knowing less? Not at all. Knowing what a tool is for might impede our ability to use it in other ways, but not knowing what it's for impedes our ability to use it at all. Knowing an inefficient solution to a problem might impede our search for more efficient ones, but solving a problem inefficiently is better than having no way to solve it. Knowledge also serves as the foundation for further knowledge. Simple ideas can support the development of more sophisticated ones, and coarse strategies can support the search for more refined ones. Education expands knowledge, and knowledge expands imagination.

In sum, this book will chart the development of imagination by describing its humble beginnings across several domains of thought and the ways it can be expanded through education and reflection. The focus is developmental, but the implications are universal. When contemplating what could happen, children are unduly constrained by their knowledge of what has happened and their expectations of what should happen—but so are adults. We are all prone to mistaking the way things are for the way they must be. But understanding why we make this mistake can help us fixate less on the ordinary and set our sights on the extraordinary.

Part I

Expanding Imagination by Example

2 Testimony

Expanding Our Historical Imagination

Pop quiz: which of the following are true?

1. Electrons are smaller than atoms.
2. A baby's sex is decided by the father's genes.
3. Human beings, as we know them today, developed from earlier species of animals.
4. Antibiotics kill viruses as well as bacteria.
5. The sun goes around the earth.
6. All radioactivity is manmade.

These statements come from a test used by the US National Science Foundation to measure public understanding of science. The first three are true, and the last three are false. Regardless of how well you did, how many of your answers were based on personal experience? Atoms, genes, viruses, radioactivity, evolution, and planetary motion are not the kind of thing we can observe first-hand. Our knowledge of what they are and how they work is based on information other people have told us. Indeed, most of our knowledge of how the world works comes not from firsthand observation but from secondhand reports—the testimony of others. Few people have orbited the earth or performed surgery, but all of us can learn about astronomy and anatomy from those who have.

The word "testimony" is most closely associated with courtrooms and eyewitnesses. We think of it as evidence for or against a disputed claim, but testimony has a much broader scope. Almost everything you learn in school comes from testimony. The typical high schooler takes algebra, geometry, biology, chemistry, physics, English, history, and government. How much of that instruction involves firsthand observation? Students may complete some laboratory exercises in science or take some field trips in social studies, but the vast majority of our school curriculum comes from books and lectures. We are told about people who lived long ago and the important contributions they made to human knowledge: Pythagoras and his theorem, Newton and his laws, Darwin and his theory, Napoleon and his wars, Shakespeare and his plays. We devote most of our childhood and adolescence to learning the collective wisdom of those who've come before us.

Testimony undergirds our personal knowledge of the world as well. When your uncle Robert tells you he grew up in Dayton, Ohio, before moving to Los Angeles to attend the University of Southern California, you believe him. But what evidence do you have that any of those claims are true? Has he shown you his diploma from USC? Has he shown you pictures of his house in Dayton? Perhaps his name isn't really Robert. Have you seen his birth certificate? And perhaps Dayton isn't a real place. Have you been there?

Most of what we're told we accept at face value. It would be impossible to learn from testimony if we questioned every statement. But sometimes we do reject testimony, and those rejections tell us something important about the relation between knowledge and imagination. All testimony can be thought of as an exercise in imagination. When we're told of something we haven't personally experienced, we model that idea in our minds, in relation to the entities and events we have experienced. Learning that your uncle grew up in Dayton draws on your knowledge of cities, the midwestern United States, and childhood. You may never have been to Dayton, but you can guess what it might be like based on your knowledge of other cities and other regions of the country. You may not have observed your uncle's childhood, but you can get a rough idea of what that experience was like from what you know about him, the time period, and children in general.

We tend not to reject a claim like "I grew up in Dayton" because this claim is easily modeled in relation to prior knowledge and easily assimilated therein. A claim like "the earth goes around the sun," on the other hand, defies prior knowledge and requires a much greater stretch of imagination. The earth feels stable under our feet, and the sun appears to move around us. The idea that the earth orbits the sun is so far removed from prior observations that it's easier to reject the idea than to forge a pathway between it and what we know. Sure enough, a quarter of Americans reject this idea, clinging instead to the belief that the sun orbits the earth.[1]

Whether we accept a claim as true depends on its source as well as its content. We sometimes accept extraordinary claims from people we trust and reject mundane claims from people we distrust. We typically think of trust as something established slowly, through a history of mutual cooperation, but it can also be discerned from more immediate cues, such as how a person speaks and whether they're in a position to know what they're talking about. From early childhood, we scrutinize the informants around us for evidence we can trust them.[2] We attend to how reliable they have been in the past, how knowledgeable they are about the current topic, and how confident they are in their assertions. Informants with such qualities tend to be trusted, but these are not the only qualities we attend to. We also consider how attractive they are, whether they resemble us in age and gender, and whether they speak with a foreign accent—qualities that have no bearing on the veracity of their testimony.[3]

For all the scrutinizing we do, valid or invalid, the content of an informant's claims can overshadow their credentials if that content is difficult to reconcile with prior knowledge. The most reliable and confident informant would be hard-pressed to convince the average American that the universe began with a huge explosion or that humans evolved from nonhuman ancestors. These ideas are often so foreign to our understanding of the world and our place within it that we find them inconceivable. The knowledge we've acquired from a few decades of personal experience prevents us from leveraging centuries of systematic observation and experimentation by others. Testimony is a gift. We're the only species that can share it and learn from it. But prior knowledge places strong constraints on the testimony we accept as true, especially in childhood.

This chapter focuses on our acceptance of entities and events that we have not personally experienced but learn about from others. Such snippets of testimony, if accepted, expand imagination by showing us that the space of real-world possibilities is larger than we had anticipated. But testimony does more than just highlight isolated facts. It underlies the learning discussed in every chapter of this book. The scientific discoveries that expand our empirical imagination, the ethical principles that expand moral imagination, and the fictional stories that expand our social imagination are all conveyed through testimony. Learning a new fact is just the tip of the iceberg.

Gullible Skeptics

There's a popular myth that children are highly credulous and willing to believe anything an adult tells them. Some scholars have argued that credulity is adaptative; evolution made children credulous so they could learn more effectively. For instance, evolutionary biologist Richard Dawkins asserts that "a human child is shaped by evolution to soak up the culture of her people. When you are pre-programmed to absorb useful information at a high rate, it is hard to shut out pernicious or damaging information at the same time. With so many mindbytes to be downloaded, so many mental codons to be duplicated, it is no wonder that child brains are gullible, open to almost any suggestion, vulnerable to subversion, easy prey to Moonies, Scientologists, and nuns. Like immune-deficient patients, children are wide open to mental infections that adults might brush off without effort."[4]

Dawkins notes that children have difficulty distinguishing fantasy from reality. He explains how a six-year-old he knows "believes that Thomas the Tank Engine really exists. She believes in Father Christmas, and when she grows up her ambition is to be a tooth fairy. . . . If you tell her about witches changing princes into frogs she will believe you. If you tell her that bad children roast forever in hell she will have nightmares."[5] Dawkins' argument is directed against religion, which he sees as no different from fairy tales, but his point extends beyond religion: children are susceptible to false information because they are naturally gullible.

Psychologists have looked into the matter and confirmed that children do, in fact, believe many things adults dismiss as impossible. In the United States, eight of every ten preschoolers believe in Santa Claus, seven believe in the Tooth Fairy, five believe in the Easter Bunny, four believe in fairies and ghosts, and three believe in dragons, witches, and monsters.[6] Belief in Santa is particularly pervasive. Children believe in Santa until age seven or eight, and they believe regardless of whether they are predisposed toward make-believe or not.[7] Children whose parents encourage belief in Santa believe more strongly than those with less-encouraging parents, but even the latter still believe. Popular culture suggests that

Santa is real, particularly around Christmas time, and children happily embrace the suggestion. Even Jewish children tend to believe in Santa, despite the absence of parental encouragement and lack of Christmas presents.[8]

Children also tend to believe in magic. If preschoolers are shown an illusion where a physical object appears to float in the air or pass through another object, they label the event as magic—"real magic"—and they bristle at the suggestion that it may have been a trick.[9] They saw it with their own eyes, after all. School-age children are better at identifying illusions as tricks, but even they claim there are real magicians in the world who perform real magic. The same has been found when children are shown physical transformations they cannot explain, such as color-changing foam or magnetic-induced levitation. Before viewing the transformations, young children claim they are impossible; after viewing them, they claim to have seen magic.[10]

But belief in Santa and belief in magic are not the full story. When psychologists have brought children into the laboratory and told them stories about physically impossible events, the children readily acknowledge that such events cannot occur.[11] Children as young as three claim that a person could move an object with their mind in a story but not in the real world. They recognize that objects cannot spontaneously appear or disappear and cannot spontaneously change shape or location. They also recognize that the laws of nature are immutable and deny that a person could grow younger, grow smaller, stay awake forever, lay eggs, turn into ice, turn into ooze, walk through a wall, walk on the ceiling, or float in the air—even if that person really wanted to or tried really hard.

The ability to distinguish possible events from impossible ones may be in place before we can articulate such thoughts. Infants who are shown physically impossible events, brought about by sleight of hand or a trick apparatus, are surprised by these events. Their surprise is revealed by an experimental methodology called "violation of expectation," where researchers determine what expectations infants hold by showing them impossible events and measuring how long they look relative to ordinary events.[12] For instance, we know that infants expect objects to move in a straight path because they look extra-long when an object changes course on its own. We know that infants expect objects to stop on contact with other objects because they look extra-long when an object seemingly passes through a barrier. And we know that infants expect objects to remain where they are put because they look extra-long when an object spontaneously disappears and reappears in a new location.

When infants stare longer at unexpected events, they probably aren't reflecting on the mutability of physical laws. They may just be noting something unusual, similar to when an adult pauses to look at a full moon or a passing eagle. But infants do other things in violation-of-expectation experiments that suggest their expectations about physical causality run deep. When six-month-olds see an impossible event, they look longer not only at the event but also at the parent who brought them to the experiment. They look at their parent's face to check to see whether they too are surprised that an object just vanished into thin air.[13] When eleven-month-olds see an impossible event, they not only look with surprise but also try to discover how it occurred. They'll pick up a car that seemed to have passed through a wall and bang it on the ground, testing its solidity; or they'll pick up a ball that seemed to float in the air and drop it, testing its gravity.[14]

All these findings, when considered together, paint a seemingly incoherent picture of children's understanding of physical possibility. Children believe in fantasy characters and magical events that defy the laws of nature while also balking at the idea that the laws of nature could be violated in real life. Intrigued by this tension, my colleagues and I have looked more closely at *how* children decide whether something is possible—the strategies they use for differentiating possible events from impossible ones. One strategy would be to compare the event under consideration with known physical principles. A child told of a person who walks through walls might consult her knowledge of physical objects and realize that walking through a wall violates the principle of solidity. Young children don't know the words "solid" or "solidity," but they do know that objects are hard and that hard things collide on contact.

This strategy seems unlikely for several reasons. It assumes that preschoolers know many different physical laws, from those that preclude floating in the air to those that preclude staying awake forever, and that they are able to access this knowledge at will. Although preschoolers may expect people to fall to the ground when they jump or fall asleep when they've stayed awake for many hours, these expectations are probably not encoded as explicit facts in the child's mind, accessible to introspection. Children may recognize that law-violating events are impossible, but they don't necessarily know why. If they did, their belief in magic would be even more mysterious because they should judge magical beings and magical events as uniformly impossible.

A more plausible strategy is that children use expectation as a proxy for possibility: any event that violates their expectations is deemed impossible, at least at first. This strategy wouldn't require identifying the physical principles that preclude an event from happening; children could simply rely on their gut, reasoning that if it doesn't happen under normal circumstances, then it can't happen under any circumstances.

If children rely on such a strategy, then they should make some peculiar errors. They should conflate events that are truly impossible with those that are merely improbable. People don't walk through walls in real life, but neither do they walk on telephone wires. Both events violate our expectations, but only one violates physical laws. Pushing this logic further, children should conflate physically anomalous events with socially anomalous ones. Just

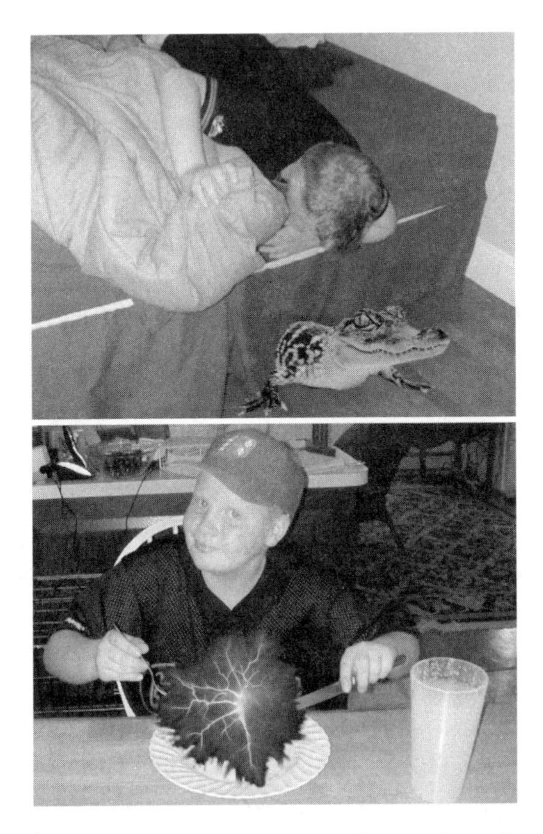

Figure 2.1 Young children deny the possibility of events that violate their conceptual expectations, whether those events are truly impossible (like eating lightning for dinner) or merely improbable (like finding an alligator under the bed). (Courtesy of the author)

as people don't walk through walls, they don't walk through the streets naked. The reasons these events do not occur are quite different, but children might deem both impossible if they fail to identify those reasons. Indeed, when asked why something is impossible, children should offer explanations that are largely irrelevant—explanations that address the question of why the event is unusual rather than why it cannot occur.

Sure enough, children make all these errors. They conflate improbable events with impossible events, judging both as impossible. They judge physically improbable events as impossible, as well as socially improbable ones. And they justify their judgments with considerations about why the events defy imagination rather than why they defy physical laws.[15]

These findings come from studies where children are read stories that contain a variety of events—some ordinary, some impossible, and some improbable. Ordinary events include activities like eating an apple or wearing a baseball cap. Impossible events include activities like walking on water or walking through a wall, which violate physical laws. And improbable events include activities like owning a lion for a pet or building a house in the shape of a teacup, which violate regularities but no laws. Such events tend not to occur but could, given the right circumstances. After the story, children are asked whether they've experienced each event and, if not, whether the event could happen in real life. If children claim the event could not happen, they're asked why not.

When adults complete this task, they claim not to have experienced the improbable events but still judge them possible. And when they judge (correctly) that law-violating events are impossible, they provide principled reasons for their judgments, citing facts about the world that preclude the events from occurring. An adult might explain that walking on water is impossible because water cannot support a person's weight or because water is not solid.

Children respond quite differently. While they too deny experiencing improbable events, they claim these events could not occur in the real world, conflating improbability with impossibility. When explaining why events are impossible, they do not cite facts or principles but instead report on what would happen if you attempted the event or how you could achieve a similar outcome by different means. For instance, children explain that walking on water is impossible because "if you tried, you'd get your feet wet" or because "you could take a boat instead"; neither explanation actually addresses the impossibility of walking on water. Very young children typically provide no explanation at all; they just reiterate their judgment, as in "a person can't walk on water" or "walking on water isn't real."

When children explain why improbable events could not occur in real life, their justifications are similar to those provided for impossible events. They claim a person couldn't own a lion for a pet because it might bite you or because you could own a cat instead. These justifications imply that children are not searching for principled, lawlike reasons why the event cannot occur. If they did, they would realize that no such reasons exist. Instead, they attempt to imagine how the event might occur, given what they know about lions and pets, and report on the mental roadblocks that pop into mind: lions are too big, they live far away, they eat other animals, they need a lot of space, they make a lot of noise, and so forth. These are reasons why owning a lion would be difficult but not reasons why it would be impossible, and children must learn to distinguish the two. Indeed, as children begin to distinguish improbable events from impossible ones, they also begin to provide better justifications for their judgments. Searching for a reason why an event is impossible may yield the realization that it's possible after all.

Children's tendency to treat improbable events as impossible spans all kinds of events, from physical to biological to psychological.[16] Preschoolers claim that walking on a telephone wire is just as impossible as walking on water, that growing a beard to your toes is just as impossible as growing back into a baby, and that reading someone's lips is just as impossible reading someone's thoughts. Knowledge of biological principles and psychological principles develops later than knowledge of physical principles, but this difference does not affect children's evaluation of possibility, presumably because they do not search for principled reasons to support their judgments. They judge expectation-violating events to be impossible by default.

Preschoolers conflate improbable events with impossible events even when they complete tasks intended to remind them that improbable events can in fact occur.[17] In one such task, four-year-olds are shown a box containing ten blue marbles and one red marble. They count the marbles of each color and then watch the experimenter dump them into an empty bag. The experimenter then retrieves a single marble and asks, without revealing the marble's color, "Could the marble in my hand be blue? Could it be red? Could it be yellow?"

Four-year-olds readily acknowledge that the marble could be blue or red but not yellow, showing they understand that "could" refers to possibility, not probability. If they interpreted "Could it be red?" as "Is it probably red?" then they would have said no. Four-year-olds are no stranger to the language of possibility; they use modal verbs like *could, can, may, must,* and *might* in their speech, and they understand that these verbs express claims of possibility (how things could be) rather than factuality (how things are).[18] The marbles task demonstrates that children understand the proper meaning of "could," but even after succeeding in this task, they still claim that unexpected events like owning a lion for a pet or growing a beard to your toes are impossible.

Strong Claims, Weak Resolve

There are actually two routes to accepting the possibility of an improbable event: you could determine that the event violates no physical laws, or you could devise circumstances under which the event might occur. Using the latter route, you could confirm that owning a lion for a pet is possible by imagining how a person might procure a lion and then care for it. Some adults prefer this strategy for assessing possibility, as we'll discuss later, but it's not a strategy children tend to use. They neither seek reasons why an event could not occur nor try to imagine how it might occur because they seem unaware of the strength of their claim.[19] Children jump from the observation that an event *does not* happen to the conclusion that it *could not* happen. But this conclusion requires justification. If something is impossible, then it will never happen—anywhere, at any time, under any circumstance. It is absent from the space of real-world possibilities.

Given how little of the world children know, it's odd they make such strong claims. A more defensible position would be to plead ignorance. But as strong as children's judgments are epistemologically, they are not that strong psychologically. They are a gut reaction, lacking grounding or backing, which makes them vulnerable to social pressure. A child who has never heard of Santa would balk at the suggestion that Christmas presents are delivered by a man who flies around the world in a reindeer-drawn sleigh; most children don't think it's possible to own a reindeer for a pet, let alone a flying reindeer. But children *have* heard of Santa. His story is widespread, and his existence is widely accepted. Parents seem to believe in him. Christmas songs celebrate him. And Christmas stories revolve around him. Isn't that evidence enough?

The conundrum of why children believe in so many impossible things is resolved not by assuming that they are credulous but by recognizing that their skepticism is shallow. Without identifying why unexpected events are impossible, children can be persuaded to accept them for social reasons, like the word of a trusted authority or the consensus of one's community. If you don't stand for something, you'll fall for anything.

Children rarely invent the fantastical ideas they are notorious for believing. These ideas are fed to them by popular culture.[20] Culture provides not only the details of what to believe—a flying man who delivers presents, a giant bunny who delivers eggs, a nocturnal fairy who collects teeth—but also the social impetus for maintaining belief. This impetus is sufficient to convince a child who judges all extraordinary events impossible but has no idea why. It's not sufficient, however, to convince a child who has begun to identify the reasons why impossible events do not occur and can use those reasons to differentiate the truly impossible from the merely improbable.

Indeed, the better children are at differentiating improbable events from impossible events, the more skeptical they are of Santa.[21] They begin to engage with the mythology surrounding Santa at a conceptual level, questioning how he could perform the feats he is said to perform. All children are curious about this mysterious gift giver, but the nature of that curiosity changes, as documented by the questions they ask about Santa. Young children want to know how cold it is where Santa lives, what his elves like to eat, and where he met Mrs. Claus, whereas older children want to know how Santa travels around the world in a single night, how he fits down chimneys, and how he knows whether each child has been naughty or nice.

As children come to appreciate the physical laws that prevent people from flying through the air or visiting millions of homes in one night, they want to know how Santa circumvents these laws. Some begin asking pointed questions. Some begin scrutinizing the evidence for Santa and his involvement in Christmas. And some begin listening to skeptical voices that were always present but never heard.[22] "Yes, Virginia, there is a Santa Claus" is the kind of thing people say only if there's doubt.

Belief in fantastical beings thus wanes in middle childhood. The prevalence of such beliefs in early childhood indicates that young children have yet to establish a firm grasp on physical possibility, but it does not mean they are willing to believe anything. They'll suspend disbelief if persuaded to do so, but disbelief is still their default.[23]

Failures of Imagination

Children's widespread but shallow skepticism tells us something important about the limits of imagination: that we are prone to make an error described by philosopher Dan Dennett as "mistaking failures of imagination for insights into necessity."[24] That is, we are prone to mistake an inability to imagine how an event might occur for evidence that it could never occur.

This error dominates the reasoning of preschoolers, regardless of context or instruction. Preschoolers deny that people can do improbable things regardless of whether they encounter the event in a book, on the internet, or in a conversation.[25] Preschoolers are more likely to accept improbable events if framed as hearsay, as in "someone told me they drank onion juice," than if framed as eyewitness testimony, as in "I saw someone drink onion juice."[26] But their default judgment, for both framings, is to claim that improbable events cannot occur.

Prompting preschoolers to imagine improbable events by asking them to close their eyes and envision the events does not help either.[27] Preschoolers reject that a person could drink onion juice even after successfully envisioning it. Their rejection of improbable events stems not from their understanding of the events but their ability to connect them to real-world circumstances. Preschoolers, like adults, can imagine hypothetical events without believing they could actually occur. Belief requires an additional layer of imagination: connecting what is claimed to be true with what we already know to be true. A preschooler might know enough about onions to realize they can be juiced and enough about people to realize they sometimes consume disgusting things, but they still side with their gut and judge the event impossible.

This disconnect—between knowing enough to judge an event possible and actually judging it possible—is made plain by asking preschoolers to explain improbable events. If preschoolers are shown a photo of a woman with a peacock and asked to explain how she came to own this bird as a pet, they can provide perfectly sensible explanations. They say she bought the peacock at a store or won it at a fair. But when the same children are asked whether a person could own a pet peacock in real life, they say no.[28] They don't find their own explanations convincing. Granted, these explanations have holes. They don't explain why a store would sell peacocks or why a fair would give away peacocks as prizes. But children do not attempt to plug these holes, nor do they attempt to decide whether owning a peacock violates any physical laws. The event is sufficiently surprising that children treat the surprise itself as evidence of impossibility.

Children's reluctance to accept what others tell them extends beyond contemplating the possibility of hypothetical events to accepting factual information about true events. Three-year-olds told about a novel animal tend to deny that it's real, even when assured that scientists collect them and study them.[29] Five-year-olds will accept the existence of novel animals but only if those animals conform to their expectations of how animals behave. Animals with unusual traits, like monkeys who live in the snow or rabbits who eat bugs, are judged to be pretend.[30] Five-year-olds are more inclined to judge unusual animals as real if a zookeeper insists they are, but a zookeeper's testimony convinces them to change their mind only about half the time.

Even ten-year-olds deny the existence of novel animals unless they are provided with information about where it lives, what it eats, or how it's related to other animals.[31] The animals a ten-year-old knows are a miniscule fraction of all the animals that exist, but this knowledge emboldens them to reject testimony about animals they haven't heard of before.

Children also reject testimony about people they've never heard of before, particularly in the context of a story. They assume that stories are not true and that the characters in those stories do not exist.[32] For instance, in one study, preschoolers were told the stories of two soldiers: a soldier who lived in Virginia and fought in the Civil War and a soldier who had a special sword that prevented him from dying in battle. When asked who is real and who is pretend, most claimed that *both* are pretend.[33] What children learn with age is not that fiction is false but that nonfiction is real.

Children's skepticism about stories extends beyond plots and characters and encompasses the very language used to tell them. When young children encounter new words in stories, such as words from a foreign language, they assume those words were made up.[34] The creators of educational shows like *Dora the Explorer* and *Little Einsteins* intend to teach children new vocabulary, and the children who watch those shows seem to learn it. They remember how the words are pronounced and what they mean. But when asked whether those words are real, they say no. "*Vámonos* means 'go,'" explained one child, "but that's made up for the show."

The same has been found for factual information presented in the form of a story.[35] While watching *Sesame Street,* for instance, children might learn that Spanish people play the claves and dance the tango, and they'll remember that information a week later. But asked whether people play the claves and dance the tango in real life, they say no. In fact, they're just as likely to sequester the factual elements of a story as the fictional ones. They deny that people play the claves in real life just as they deny that penguins play the bongos in real life. What's learned in a story stays in the story, as we'll discuss further in chapter 9. Parents need not worry that their children might think growing an extra arm is possible if they see SpongeBob SquarePants grow an extra arm, but they might worry their children will become skeptical of real objects, like intercoms and meat tenderizers, after they see SpongeBob use them.[36]

Distinguishing fact from fiction requires coordinating two types of knowledge: general knowledge of what can happen in the world

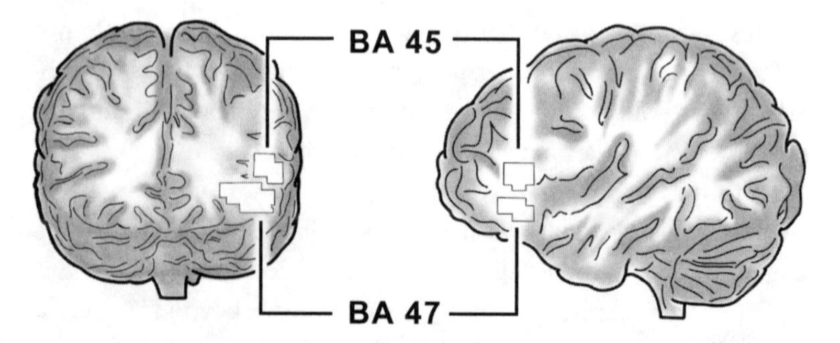

Figure 2.2 Imagining an encounter with a fictional person, like Cinderella, activates brain regions in the inferior frontal gyrus, associated with the retrieval of semantic memories. These regions are less active when imagining an encounter with a real person, like Barack Obama. (Adapted from Abraham et al. 2008) (Courtesy of the author)

and personal knowledge of what has happened. General knowledge is needed to outline the contours of a hypothetical event, whereas personal knowledge is needed to fill in the details. Brain-imaging studies have found that we rely on general knowledge and personal knowledge to different extents when contemplating fictional events or realistic ones.[37] Adults asked to contemplate a fictional event, like meeting Cinderella, show heightened activity in brain areas associated with general (semantic) knowledge, whereas adults asked to contemplate a realistic event, like meeting Barack Obama, show heightened activity in brain areas associated with personal (episodic) knowledge.

Most adults have never met Barack Obama, but we recognize we could. Not only can we imagine the meeting itself—approaching Obama, saying hello, shaking his hand—but we can imagine the circumstances under which such a meeting might transpire. Perhaps we are shopping at the same store, or we are introduced by a mutual friend. The mere fact that Obama is a real, living person assures us that such circumstances are possible because there is no physical law that precludes them. This is the logic of possibility.

The distinction between possible and impossible events rests not on present circumstances but on hypothetical ones and the laws that constrain them. Adults appreciate this logic, at least tacitly, but where does such an appreciation come from? How do children learn that actual events are a mere subset of all possible events?

One suggestive finding is that children are more likely to judge improbable events possible if they first consider the circumstances surrounding those events. Kindergartners who learn the steps for making unusually flavored ice cream, such as pickle-flavored ice cream, are more likely to accept the possibility of other unusually flavored ice creams.[38] Likewise, preschoolers who hear read a story about someone owning a lion for a pet are more likely to accept that this event could happen in real life if their parents foster speculation on how it might occur.[39]

Another way of helping children recognize the possibility of unexpected events is asking them to consider whether those events could occur in a distant land.[40] Children who deny the possibility of drinking onion juice will nonetheless concede that someone "in a country very far away" could do it. People in distant countries are also given the benefit of the doubt as to whether they can ride hippopotamuses, grow beards to their toes, build houses from toothpicks, or own zebras for pets. When such events are described as occurring far away, children either suspend the expectations that would lead them to deny the event or they hold those expectations in abeyance, reasoning that the world might operate differently somewhere else.

Prompting thoughts of distant lands has consequences for adults as well. Adults, like children, believe that improbable events are more likely to occur somewhere else.[41] We think rare genetic conditions are more common on the other side of the country, that underdogs are more likely to win a sporting event in another city, and that gamblers are more likely to get a lucky hand in a faraway casino. Distant events are thought of more abstractly than local events, which leads us to relax the specific, concrete beliefs that guide our expectations about an event's likelihood.[42]

Ignoring improbable events is a practical solution to the problem of deciding what events we should plan for and what events we can ignore. But it's not always an optimal solution, particularly if we end up rounding the probability of an improbable event all the way down to zero. Mistaking improbability for impossibility leaves little room for learning something unexpected, as discussed in chapter 4, and even less room for creating something unexpected, as discussed in chapter 3.

Social Improbabilities

In 2009, the candy company Starburst ran a series of commercials to emphasize that their product is "solid yet juicy," a seeming contradiction. In one commercial, a Korean man sees his son eating a Starburst and exclaims—in a Scottish accent—"Look at this! One contradiction eating another." The son looks puzzled, so the father explains: "You're Scotch-Korean, and Starburst is a solid, yet juicy like a liquid." The son then smiles and says, "Contradictions taste good."

The idea that a person of Korean ancestry could live in Scotland and speak with a Scottish accent is not a contradiction. Hundreds of Scottish citizens fit that description. Still, the creators of the Starburst commercial see the combination of Scotch and Korean identities as so unusual as to stretch imagination. In fact, in a follow-up commercial, they liken this combination to a truly impossible one: the living dead. The Scotch-Korean man encounters a zombie, and the two argue over whose combination of identities is more contradictory.

These commercials are an extreme example of a more general tendency to balk at unfamiliar social possibilities. The social world is characterized by regularities, just like the physical world, and we are surprised by people and situations that violate them. While adults generally accept that social anomalies are possible, young children reject those anomalies, classifying them as impossible just like physical anomalies. They claim it's not possible to alter traditions, customs, procedures, cultural associations, object labels, rules of etiquette, or gender roles.[43]

For instance, they deny that a child could sing jingle bells at a birthday party, wear pajamas to the grocery store, or wear a bathing suit to school. They deny that adults could get together and change the name of dogs to wugs, change the color of stoplights from red to purple, or change the side of the road we drive on. They claim it's not possible to eat food with your hands, take a bath with your shoes on, or ask for something without saying "please."

These judgments are not absolute. Young children do show some recognition that social regularities are more mutable than physical laws. When explaining why people conform to social regularities, they cite reasons rather than causes—desires and permissions rather than capacities and capabilities.[44] When asked whether anomalous events could occur on another planet, they agree that social anomalies are more likely to occur than physical ones, conceding that the citizens of another planet might call dogs wugs even if they can't make rocks float in water. Children are also more accepting of social anomalies if asked whether they could happen in another country or in the distant future.[45] Children's understanding of social norms also varies across cultures, with children from Western cultures denying that a person could violate social norms less adamantly than children from Eastern cultures.[46]

Children do, then, appreciate that social norms are not as restrictive as physical laws, but this appreciation is largely implicit. It manifests itself in explanations or thought experiments and typically only when children are asked to consider several anomalous events. If you ask preschoolers whether a particular social anomaly is possible, most will say no. They claim it has not occurred in the past and will not occur in the future, no matter who is involved or why.[47]

One concern with these findings is that children who claim a social anomaly couldn't happen really mean that it shouldn't—that it is impermissible. But this confusion is part of the point. If children's understanding of what could happen is grounded in what they expect to happen, then they should confuse possibility with permissibility, at least before they've learned to reflect on their expectations.

My colleague Jonathan Phillips and I explored whether children truly conflate possibility with permissibility by asking them to make both judgments for the same events.[48] We presented preschoolers and elementary schoolers with a variety of unexpected events. Some violated moral rules, like stealing candy; some violated social conventions, like wearing pajamas to school; some violated physical laws, like floating in the air; and some violated empirical regularities, like seeing a movie for free. For all types of violations, we asked children whether the event could happen in the real world and whether the event was okay to do.

We found that, by age eight, children differentiated the questions and the violations. Asked about possibility, they claimed a person could not violate physical laws but could violate moral rules, social conventions, and empirical regularities. Asked about permissibility, they claimed it would be wrong to violate moral rules but not wrong (or as wrong) to violate social conventions, empirical regularities, and physical laws. Four-year-olds, on the other hand, claimed it was both impossible and impermissible to commit any of these violations. They claimed that floating in the air is not just impossible but also wrong and that stealing candy is not just wrong but also impossible. Even minor violations, like wearing pajamas to school or seeing a movie for free, were judged harshly; preschoolers claimed these event neither could happen nor should.

The idea that children are rigid in their expectations about what can and cannot happen has been noted by many psychologists, including one of the first developmental psychologists, Jean Piaget. In his treatise on moral development, Piaget noted that "[social] rules are naturally placed by the child on the same plane as actual physical phenomena. One must go to bed at night, have a bath before going to bed, and so forth, exactly as the sun shines by day and the moon by night, or as pebbles sink while boats remain afloat. All these things are and must be so; they are as the World-Order decrees that they should be, and there must be a reason for it all."[49] Differences between physical and social possibility are not innately intuited; they must be learned. And even after we learn them, the two forms of possibility remain linked in our mind.

My colleagues and I explored this link in a study with adults, where we compared adults' judgments of physical possibility to their judgments of moral permissibility.[50] The relevant considerations, for both judgments, are circumstances that would allow the events to occur or principles that would preclude them from occurring. Adults recognize this logic, at least implicitly, but rely on one set of considerations more than the other.

Some adults rely mainly on circumstances. They have expansive views of what's possible and what's permissible. They claim that physically extraordinary events, like teleportation and intergalactic travel, are possible, and they claim that socially extraordinary actions, like flag desecration and consensual incest, are permissible. When asked to justify their judgments, they appeal to mitigating circumstances. Teleportation is possible if the object is super-small; intergalactic travel is possible if one's spacecraft is super-fast; flag desecration is permissible if done in private; and consensual incest is permissible if safeguards are taken against pregnancy.

Other adults rely mainly on principles. These adults have more restrictive views of what's possible and what's permissible. They tend to deny the possibility of extraordinary events, as well as the permissibility of extraordinary actions, and they justify their judgments by appealing to violations of general principles. Teleportation violates the conservation of mass; intergalactic travel violates the conservation of energy; flag desecration violates federal law; and consensual incest violates biological imperatives against inbreeding.

Because questions about possibility and permissibility are hypothetical, not factual, it's difficult to say whether reliance on circumstances is better or worse than reliance on principles. Still, the fact that adults' reasoning about social possibility mirrors their reasoning about physical possibility implies that the two forms of reasoning remain interconnected beyond childhood. Adults recognize that the question of whether something *could* happen is different from the question of whether it *should* happen, but we draw on the same considerations—circumstances or principles—for answering both.

One reason adults differ in whether they rely on circumstances or principles is that they may see these considerations modeled to different degrees in their preferred genre of fiction. People who read a lot of science fiction are more accepting of the possibility of extraordinary events as well as the permissibility of extraordinary actions. People who read romance novels, on the other hand, are more likely to reject both.[51] While romance novels reinforce traditional notions of morality and causality, science fiction challenges them. In fact, when people watch science fiction shows, such as the space western *Firefly* or the futuristic police drama *Continuum,* they adopt a more expansive view of permissibility, at least in the moment.[52] Science fiction primes our ability to generate circumstances that might render a moral transgression acceptable. For example, most people agree that kidnapping a child is wrong, but is it always wrong? What if the child is being mistreated? What if the child is in danger? What if the parents are in danger from the child?

Science fiction provides a workspace for exploring the malleability of real-world causes and constraints, as we'll discuss in chapter 9. This kind of thinking is a prerequisite for imagining dramatic changes to the status quo, physical or social. Indeed, social changes can be even more difficult to imagine than physical ones. Consider our reaction to hearing of an atrocity, like a school shooting or a suicide bombing. The first question we ask ourselves is not "Why did this happen?" but "How could this have happened?"—a question about possibility. We know, reflectively, that moral rules can be broken, but we don't expect them to be broken, and when they are, we fixate on the possibility of the transgression before pondering the means and motives of the transgressors. Social atrocities mentally transport us to a new location in the landscape of possible events because they lie far from the path of expected events. When testimony expands our sense of possibility, it's not always a pleasant experience.

3 Tools

Expanding Our Technological Imagination

Traveling by air was a fantasy for most of human history—a notion as whimsical as traveling to the center of the earth or traveling back in time.[1] The mathematician Simon Newcomb declared that "no possible combination of known substances, known forms of machinery, and known forms of force can be united in a practical machine by which man shall fly long distances through the air." The physicist Lord Kelvin told fellow scientists, "I can state flatly that heavier-than-air flying machines are impossible." Less than ten years later, the Wright Brothers had flown the first powered aircraft. Air travel instantly went from an impossibility to an actuality.

From our modern, jet-setting vantage point, denouncements of air travel sound silly, even hubristic. But we too would join the chorus if we hadn't seen airplanes cross the sky or traveled in one ourselves. In fact, children growing up in today's highly technological society are as skeptical as Lord Kelvin when it comes to technologies they have not personally observed.

Children as old as six deny the possibility of a machine that makes your voice sound like somebody else's or a machine that tells you if there's metal nearby, even though they could purchase both at a local store.[2] They reject such machines nearly as often as they reject machines that violate physical laws, such as a machine that makes a heavy toy float in the air or a machine that makes a grownup into a child. Voice changers and metal detectors are deemed as impossible as levitators and age reversers if we haven't encountered them. Indeed, readers of the future may scoff at the suggestion that levitators and age reversers are impossible, as they read this book in their hovercraft on their thousandth birthday.

The science fiction writer Arthur C. Clark famously quipped that "any sufficiently advanced technology is indistinguishable from magic."[3] This quip captures not only our amazement at new technologies but also our ignorance of the mechanisms of familiar ones.

Consider the helicopter. You've probably seen so many helicopters that you wouldn't stop to marvel at the next one you saw, but do you know how helicopters work? People who are asked to rate their knowledge of helicopters recognize that they know less than a pilot or a mechanic, but they still think they know a fair amount, rating their understanding as a four on a seven-point scale.[4] If a rating of four sounds reasonable, stop for a moment and try to explain helicopter flight. Do you know anything about the controls in a helicopter's cockpit or the configuration of its gears? Do you know what forces allow a helicopter to hover in the air, let alone move forward?

When people are asked to rate their knowledge of mechanical devices like helicopters, their ratings reflect knowledge they don't actually possess. Their initial rating of a four drops to a three after they try to explain how the device works and then drops even lower after they are asked a pointed question like "How does a helicopter move forward?"[5] This pattern has been observed for many devices, from sewing machines to toilets to zippers, and under many conditions, including when people are warned in advance they'll be asked to provide an explanation. It emerges early in childhood and does not abate with expertise.[6] Experts obviously know more than novices, but they too discover they know less than they think they do when prompted to provide explanations.

Even a device as simple as a bicycle remains a mystery for people who ride one every day.[7] In one study, people were asked to rate their knowledge of how a bicycle works before drawing one from memory. While most people rated their knowledge as a four out of seven, their drawings included a number of egregious errors, from affixing the pedal to the back wheel, rendering the gear chain useless, to placing the chain over both wheels, rendering the front wheel unsteerable. Following the drawing task, people were provided with pre-made drawings, some correct and some incorrect. Their ability to identify the correct drawings was no better than their ability to draw a bicycle themselves. Professional cyclists were generally more accurate than amateurs, but even the professionals made mistakes.

Drawing a bicycle sounds easy until you try. The vividness of our mental image, combined with the salience of our past experience, creates an illusion of understanding. We mistake our ability to imagine a bicycle for evidence that we understand how bicycles work. It is the converse of the mistake made by Lord Kelvin, who mistook his inability to imagine how an airplane could work for evidence that airplanes are impossible. This conflation of imagination and understanding highlights the difficulty of learning by example. Technology, like testimony, expands our sense of possibility by transporting us to a new location in the landscape of relevant ideas, but it's hard to connect that location to locations known through prior experience. And it's even harder to discover new locations on our own, by inventing something novel.

WHICH SHOWS THE USUAL POSITION OF THE FRAME

WHICH SHOWS THE USUAL POSITION OF THE PEDAL

WHICH SHOWS THE USUAL POSITION OF THE CHAIN

Figure 3.1 Familiarity with a technology is no guarantee of understanding how it works. People who ride a bicycle every day often fail to know how the frame is structured (top left), where the pedals are located (middle left), and how the chain is attached (bottom right). (Adapted from Lawson 2006) (Courtesy of the author)

Invention Apprehension

The long view of technology, from stone tools to power tools, suggests that humankind is consistently marching toward progress, but the real history of innovation is a series of fits and starts. "At first people refuse to believe that a strange new thing can be done," notes the author Frances Hodgson Burnett. "Then they begin to hope it can be done, then they see it can be done—then it is done and all the world wonders why it was not done centuries ago."[8] New technologies have radically transformed our lives, rewriting our understanding of possibility along the way, but their development is impeded by doubt and disbelief, and their introduction is met with distrust and disdain.[9]

Consider the public's distrust and disdain of COVID-19 vaccines, developed in the hopes of ending a globally disruptive pandemic. These vaccines were shown to be safe and effective in clinical trials, but people remained concerned about their long-term effects as well as the possibility they include fetal tissue, carcinogens, or surveillance microchips.[10] Vaccine skepticism allowed the coronavirus pandemic to kill many more people than it should have, despite widespread efforts to counter vaccine disinformation.

Concerns about fetal tissue and microchips may be vaccine specific, but fixating on the potential dangers of new technology is widespread. Historically, people have rejected dozens of technologies that are now thoroughly embraced, including refrigerators, tractors, wind turbines, electricity, coffee, margarine, water purifiers, the printing press, mechanized looms, recorded music, guns, cars, and cell phones.[11]

Sometimes new technologies are rejected for economic reasons, such as when those who profit from an existing technology fear the loss of wealth and employment that will follow its replacement. Handloom weavers in nineteenth-century England were not enthusiastic about the development of a mechanized loom, which promised to save time and labor in the production of textiles, because such looms threatened to put them out of work. In protest, a weaver's guild known as the Luddites broke into textile factories and smashed the machines, forever associating the name Luddite with someone who opposes innovation.[12]

Economics is not the only reason people reject new technology. Many are rejected because people remain unduly skeptical of their safety.[13] Coffee was initially suspected of causing exhaustion, paralysis, and impotence. Margarine was suspected of causing sore eyes, brittle hair, and weak bones. Electric lights were blamed for insomnia, and refrigerators were blamed for food poisoning. Today, we happily store margarine in our refrigerators and drink coffee by the glow of electric lamps, but our skepticism has shifted to other technologies, such as vaccines and genetically modified foods. Much of this skepticism stems from ignorance of how these technologies work, which in turn fuels disregard for demonstrations of their safety.[14]

In the public's defense, some new technologies are dangerous and should not be embraced until adequately tested. Untested medical cures have historically caused more harm than good, lowering life expectancies for those who took them.[15] But safety concerns are just one manifestation of a general tendency to reject new technologies. A more pernicious reaction is to underestimate a technology's utility.[16] Tractors were initially dismissed as no more useful than horse-drawn plows, refrigerators as no more useful than ice boxes, and muskets as no more useful than crossbows. Early versions of these technologies were indeed less practical than later versions, but people consistently failed to grasp their potential. Life as we know it blinds us to life as it could be.

Consider the seemingly banal technology of artificial ice.[17] We use ice so regularly—adding it to our beverages, packing it into our coolers, applying it to our injuries—that ice makers are a standard feature of home freezers. What would life be like without ice? People used to enjoy ice only in the winter, when it was made by nature, leading to the emergence of a whole industry that harvested and stored ice for year-round use. During the nineteenth century, some innovative ice harvesters realized that there were places in the world where ice never forms, such as the Caribbean and the West Indies, and they spent vast fortunes shipping ice to these regions. But the people there were not as enthralled with ice as expected; they saw it as a curiosity with no practical value. It took decades for the ice trade to turn a profit because it took decades for people in tropical climates to reimagine how they might store and prepare food with the aid of ice.

Or consider the even more banal technology of the wheeled suitcase.[18] The baggage maker Bernard Sadow spent years trying to convince retailers to sell his product before Macy's finally agreed in the mid-1970s. Prior to Sadow, at least five people had patented designs for similar technology, with the first patent dating back to 1925. Even after Macy's began selling wheeled suitcases, they did not become popular until the advent of the vertically oriented "rollaboard." Today, virtually everyone at the airport has a rollaboard, but it took decades for luggage retailers to recognize the practical value—and hence commercial value—of sticking wheels on suitcases.

Even technologies recognized as life-altering at the time of their invention often end up altering our lives in unexpected ways.[19] Computers were heralded as time-saving number crunchers and were immediately embraced by universities and government agencies, but few foresaw the mass appeal of the computer and its utility for writing, communicating, bookkeeping, graphics making, and game playing. The same is true for the cell phone. Many saw the value in being able to make phone calls from anywhere, but few saw the value of a handheld device that allowed you to connect to the world by other means, including text messaging, social media, and GPS navigation. The ground broken by groundbreaking technologies is rarely marked in advance and, once broken, is rarely remembered as having been unmarked.

Whence Comes Innovation?

As difficult as it is to accept new technologies, it's even more difficult to invent them. Technological innovation appears to be progressing at a steady clip, but many of those innovations are superficial.[20] And many industries have hit a wall in pursuing established technological goals, such as shrinking the microchip.[21] We forget, in the age of ever-improving phones and ever-updating software, that technological innovations have historically been few and far between. You may be on your fifth iPhone, but it took humans hundreds of years to develop the first telephone and nearly a hundred more to develop the first cellphone. So what explains humans' ability to innovate but also our slowness at doing so?

Some scholars, like psychologist Alison Gopnik, point to the trade-offs between two forms of learning: exploration and exploitation.[22] Exploration is a prerequisite for innovation; innovative ideas are discovered only by searching through a broader space of possibilities. But searching can be futile if we are searching in the wrong place or with the wrong criteria. Even productive searches can be futile if they take too long; the time spent searching for a better option—a better design, a better strategy, a better theory—might have been better spent exploiting a known option.

Exploitation is the opposite of exploration; it is taking advantage of the option at hand rather than search for something new. According to Gopnik, humans balance the trade-off between exploration and exploitation on a developmental timescale; children spend more time exploring than exploiting, whereas adults do the opposite. Gopnik describes children as the research and development division of humanity and adults as the production and marketing division.[23]

This view of innovation explains both its origin and its decline: we are programmed to explore the landscape of possibilities in search of new options but abandon this impulse when we find an option we can exploit. Children are born as freewheeling innovators, but age and experience turn them into narrow-minded laborers. It's a poetic view of children, and a clever explanation of innovation, but it doesn't accord with studies of children's actual innovation, at least not in the context of tools. These studies find that children are excellent at mimicking how other people use tools, but they rarely invent their own.

Consider the problem of retrieving a toy that has fallen to the bottom of a narrow tube. The tube can't be moved, but the toy has a handle on it, and you have a pipe cleaner at your disposal long enough to reach the toy. What do you do? Most adults see the solution immediately; they bend one end of the pipe cleaner into a hook and then use the hook to fish out the toy. The solution seems so straightforward that you might object to calling it an innovation; yet for preschoolers bending a pipe cleaner into a hook is truly innovative. Fewer than 10 percent of preschoolers discover this solution on their own.[24] Instead, they stick the pipe cleaner into the tube without bending it and are unable to lift the toy.

Preschoolers fail to innovate in several other problem-solving tasks as well. When given a folded pipe cleaner and asked to retrieve a pom-pom from inside a horizontal tube, most preschoolers fail to realize they can unfold the pipe cleaner and use it as a prod.[25] When given some yarn and asked to retrieve an object on the other side of a fence, most preschoolers fail to realize they can loop the yarn around the object and pull it toward them.[26]

In one far-reaching study, preschoolers were tested on their ability to innovate in twelve situations—situations analogous to those in which chimpanzees and orangutans have been observed to innovate in the wild.[27] For instance, chimps and orangutans use sticks to perforate termite nests and stones to crack nuts, and preschoolers were tested on their ability to use sticks and stones in a similar manner, albeit to retrieve stickers rather than food. The preschoolers' success rate across the twelve tasks was only 32 percent, with most preschoolers failing most tasks. Human children do not reliably invent the tools that even apes use. Granted, individual apes do not reliably invent tools either; they copy the tools they see others using, shifting the burden of innovation from the individual to the group. Still, the tool-innovation skills of human children are not exceptional from the start.

Children's failure to innovate is surprisingly widespread. Preschoolers have been tested on the hook task around the globe, and they fail regardless of whether they are growing up in an industrialized society like Brisbane, Australia, or a nonindustrialized one like the Bushman settlements of South Africa.[28] They fail under conditions meant to help them succeed, such as when they are encouraged to manipulate the materials beforehand or when they are explicitly instructed to make something new.[29] And they fail even after exhibiting high levels of creativity in drawing tasks, where there is no solution per se.[30]

To succeed in the hook task, children need to recognize that the pipe cleaner is pliable, that the toy is retrievable only with a hook, and that the pipe cleaner can be turned into a hook by bending its end. Researchers have helped children achieve each of these insights and found that some are more valuable than others. Showing children that the pipe cleaner is pliable doesn't aid performance, but showing them they need a hook—by letting them play with a premade hook before giving them the pipe cleaner—is more helpful.[31] But even seeing a premade hook is not enough for many preschoolers; they also need to see a demonstration of how to bend the pipe cleaner into a hook before they will make one on their own.[32]

Older children require fewer hints to solve the task than younger children, but the fact that preschoolers do not spontaneously solve this task undermines the idea that we are natural-born innovators. Preschoolers not only need help inventing tools, they also need help using them. Although everyone may need help using complex tools with hidden affordances, like a TV remote, preschoolers also need help using simple tools with obvious affordances, like rakes and shovels.[33] When given such tools, children as old as three fail to use them to solve basic tasks like retrieving an object from the other side of a barrier or lifting an object from inside a container. They can learn quickly and effectively from a demonstration, but they flounder on their own. It's possible that young children fail tool-manipulation tasks because they are more intent on exploring the tools than exploiting them. But their exploration does not often lead to success, even after many tries. Exploration may be necessary for innovation, but it is not sufficient. Innovation requires additional skills, which young children seem to lack.

Unimaginative Imitation

Perhaps it's not a problem that children fail to innovate so long as they imitate. If someone in the community happens to innovate, by plan or by accident, children can benefit from the innovation by copying it. Children do imitate others' tool use and tool manufacture, as noted previously, but their imitation behavior has some surprising quirks. Consider children's behavior in the "floating peanut" task, where the goal is to retrieve a peanut from the bottom of a narrow tube. Children are not given a pipe cleaner to fish it out but rather a pitcher of water. The solution is to pour the water into the tube, allowing the peanut to float. With enough water, the peanut will float to the top where it can be retrieved by hand.

Preschoolers fail this task, as they fail other innovation tasks, but they can learn to succeed by watching someone else succeed.[34] That is, if they watch someone retrieve a peanut by pouring water into the tube, they will repeat that person's actions when asked to retrieve the peanut on their own. And they repeat those actions to the letter: they don't just pour water into the tube but pour it in precisely the same manner they saw someone else pour it. For instance, they will pour the water in two steps—from the pitcher into a cup and then from the cup into the tube—if they saw someone else pour the water in that manner. The insight that they can retrieve the peanut by raising the water level is accompanied by a lack of insight that the cup is unnecessary.

Might children have thought the cup was necessary for controlling water flow? Or that the pitcher was too heavy to maneuver? Probably not: children who saw an adult pour water directly from the pitcher into the tube repeated her action, foregoing the cup and still retrieving the peanut. Mimicry of this type is not restricted to the peanut task. Children have been observed to copy unnecessary or inefficient actions in many other situations, from playing with new toys, to retrieving hidden objects, to using kitchen utensils.[35] When children watch others perform a novel task, they don't just imitate; they *over*imitate.

Overimitation has been observed in a variety of contexts, though it has been studied most extensively using puzzle boxes. These boxes contain a prize on the inside that can be retrieved by manipulating affordances on the outside, such as levers, handles, and doors. Some of these affordances are functional, and others are not. In one of the first studies of overimitation, preschoolers were given transparent boxes in which the functional affordances were clearly distinct from the nonfunctional ones, such as a box with a removable handle nowhere near the compartment where the prize was stashed.[36] When children were asked to retrieve the prize on their own, almost none removed the handle; they went straight for the compartment. But when children saw an adult remove the handle before opening the compartment, they dutifully repeated this unnecessary action.

In follow-up experiments, the researchers tried to dissuade children from repeating unnecessary actions with training, warnings, and incentives.[37] They demonstrated a silly action, like tapping the box with a feather, and told children "I want you to watch really carefully, because when I open this box, I might do something that's silly and extra, just like the feather." They urged children to avoid "anything silly and extra" and "only do the things you have to do."

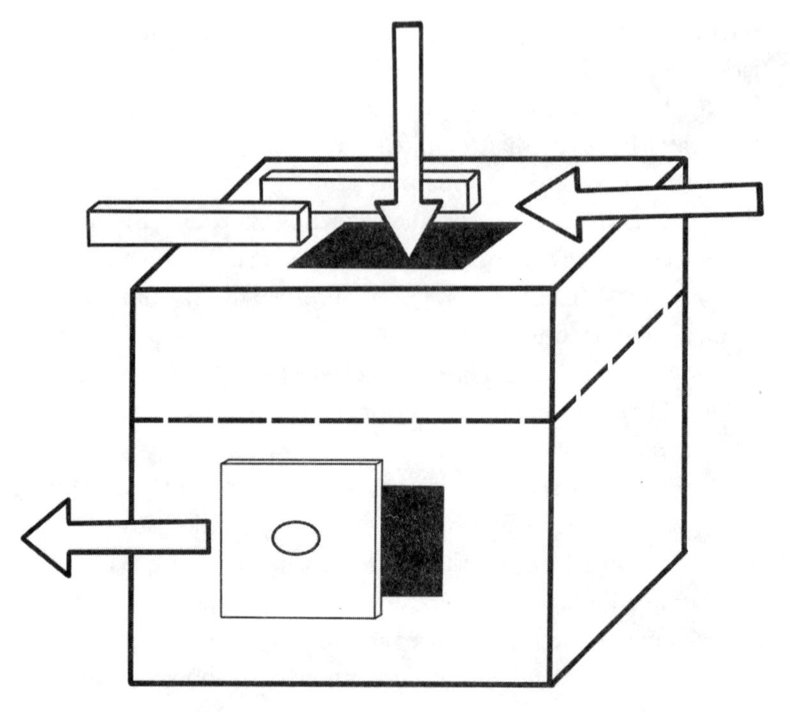

Figure 3.2 To retrieve a prize from this box, one need only open the bottom door, but young children will first manipulate irrelevant features on top of the box if they observe someone else do the same. (Adapted from McGuigan 2013) (Courtesy of the author)

They even put children under time pressure, telling them that the experiment was running behind schedule and that they needed to retrieve the toy as quickly as possible. In all cases, children continued to perform unnecessary actions—actions they wouldn't have performed if they hadn't seen them performed by others.

Following the discovery of overimitation, researchers have begun uncovering the factors that shape it. They have found that children repeat unnecessary actions if they are part of a demonstration but not if they observe those actions by happenstance.[38] Children are also more likely to repeat unnecessary actions if the consequences of those actions cannot be observed, such as pulling a lever when they can't see what the lever is connected to.[39] Children growing up in industrialized societies are more likely to overimitate than those growing up in nonindustrialized ones like Fiji and rural Peru, but even children in nonindustrialized societies repeat unnecessary actions more often than not.[40] And the children most likely to repeat unnecessary actions are not toddlers, just beginning to use tools on their own, but preschoolers and elementary schoolers. Overimitation actually increases with age.[41]

This last finding implies that overimitation is not an innate deficit but rather an emerging feature of social learning. And it appears to emerge only in humans. Researchers have looked for overimitation in other social animals, including chimpanzees, bonobos, and dogs, by placing food inside puzzle boxes and then showing the animals how to retrieve the food with a mixture of necessary and unnecessary actions. The animals tend to omit the unnecessary actions. If they do repeat them, they quickly stop after they realize the food can be retrieved more simply.[42]

So why does overimitation emerge in humans? Toddlers may be confused about how to use many tools, but it's older children who overimitate—children who have ample experience with tools and their affordances. Are they confused as well? When five-year-olds in overimitation studies are asked why they repeated an unnecessary action, most acknowledge that the action was physically unnecessary but insist that it was *socially* necessary—that it is the right way to accomplish the task.[43] In fact, when children who've been taught the "right" way to open a puzzle box watch someone else open it the wrong way—without any unnecessary actions—they protest the omission and insist that person include the omitted steps.[44]

Overimitation is thus a form of socialization: it's about conforming to norms rather than misanalysing causes. The norms are conveyed through pedagogical demonstrations and enforced through social monitoring. Children are more likely like to repeat an unnecessary action if it is demonstrated by someone they respect, such as the principal of their school, but less likely to repeat that action if the person who demonstrated it leaves the room and is no longer present to observe them.[45] Overimitation also varies with children's attentiveness to social cues; children with autism, who are generally inattentive to such cues, almost never overimitate.[46]

An extreme example of socially motivated imitation is ritual.[47] Rituals demand high imitative fidelity. They are learned from particular authorities, performed in the presence of particular peers, and associated with particular outcomes. These features are true of religious rituals, but they are also true of secular ones. Moving the tassel on a graduation cap from the right side of the cap to the left is no more efficacious than blowing a ram's horn on Rosh Hashanah or swinging burning incense at a Catholic mass. The point of the activity is social rather than causal, marking the occasion and its participants as special.

In short, humans repeat causally inert actions as a way of facilitating social bonding. Is this behavior harmless, from a problem-solving perspective, or might it hamper our ability to innovate? Studies with adults suggest it might. These studies have found that adults overimitate as much, if not more, than children.[48] Given transparent puzzle boxes, where the functions of each affordance are clear, adults remove unnecessary bolts and open unnecessary doors if they see someone else demonstrate these actions. Most adults probably realize these actions are unnecessary, just as most children do, but our knee-jerk tendency to overimitate may cloud our perception of the problem and hinder our ability to solve it more efficiently.

The psychologists Gyorgy Gergely and Gergely Csibra relay a pointed example of how overimitation can cloud our judgment, even as adults.[49] An acquaintance of theirs named Sylvia had a particular way of making ham. Before putting a ham in the oven, she sliced a chunk from both ends. One day when Sylvia's mother came to dinner, she observed Sylvia prepare a ham in this unusual manner and asked what she was doing. Sylvia was taken aback and explained that she was just repeating what her mother had taught her, to which her mother replied "But that is because I did not have a wide pan!" Sylvia was an intelligent person—a researcher by trade—but she never questioned the tool-use technique she learned from a trusted authority. She assumed the technique was optimal yet could easily have imagined a more optimal one if she had tried.

Fixating on the Familiar

Sylvia's recipe for ham illustrates more than just a blind tendency to copy others; it illustrates how readily we stick to a familiar plan or protocol, even when a better one could be found. Psychologists label this tendency *Einstellung,* which is a German word for "attitude" or "mindset." Einstellung was first documented in a study involving simple math problems.[50] Participants needed to find a way of obtaining a precise amount of water using jars of three different volumes, such as obtaining 100 quarts of water using a 127-quart jar, a 21-quart jar, and a 3-quart jar. How might you do it? The answer is to fill the 127-quart jar and then remove 21 quarts using the 21-quart jar and 6 quarts using the 3-quart jar twice.

Here's another problem: how might you obtain 5 quarts of water using a 43-quart jar, an 18-quart jar, and a 10-quart jar? The answer is to fill the 43-quart jar and then remove 18 quarts, followed by 10 quarts twice. These problems model a simple arithmetic procedure: take the volume of the large jar and then subtract the volume of the medium jar and two times the volume of the small jar.

When participants are given a series of such problems, they often fail to realize that sometimes there are shortcuts. Consider the problem of obtaining 20 quarts of water using a 49-quart jar, a 23-quart jar, and a 3-quart jar. You could begin by filling the large jar and removing water as usual (49 minus 23 minus 3 minus 3) or you could fill the medium jar and remove one small jar of water (23 minus 3). The more practice we accrue with the three-step solution, the less likely we will discover this two-step shortcut.[51]

Even experts fall prey to Einstellung. Chess players skilled enough to enter professional tournaments stick to the moves they know best, despite the availability of better moves.[52] In one study, researchers assessed expert chess players' ability to consider novel strategies by recording where on the chess board they looked while trying to find the shortest path to victory. They consistently looked at squares associated with familiar solutions rather than squares associated with more efficient solutions well within their grasp.[53]

The saying "practice makes perfect" should really be "practice makes adequate." Adequate solutions block our ability to imagine better ones. Nowhere is this problem felt more keenly than in designing new tools for solving old problems. The tools we currently use block our ability to imagine better ones. For this reason, the economist Joseph Schumpeter describes innovation as "creative destruction": new tools necessitate the destruction of older ones.[54]

We have seen historical examples of how new technologies were slow to develop because people couldn't see how to improve older ones, by, say, adding wheels to suitcases or adding cameras to phones. Psychologists have studied this shortsightedness in the laboratory by asking people to design new versions of familiar products, such as a bike rack that can be mounted on a car or a coffee cup that minimizes spills.[55] Some participants are given examples of the product they have been asked to design and others are given instructions but no examples. The group given examples would seem to have an advantage, as more information is generally better than less, but examples do more harm than good. Participants given examples stick too closely to them, generating products that include the same features, as well as the same flaws.

For instance, participants shown an example of a car-mounted bike rack that includes suction cups and tire braces tend to include those features in their own design, even when they are warned that tire mounts make the rack difficult to load and suction cups can lose their grip. Likewise, participants shown an example of a nonspill coffee cup that includes a mouthpiece and an internal straw replicate both features, despite warnings that the mouthpiece could leak and the straw would be difficult to clean. Such features almost never appear in the designs of participants given instruction but no examples. Their designs end up being more original and more versatile.

Product engineers are aware of this problem and call it "design fixation." But awareness of the problem doesn't help them avoid it. Expert engineers given examples of a target technology include the features of those examples in their own designs, just like novices. And the influence of examples is often more pernicious for experts than it is for novices, because they lead experts to adopt the example's overall structure, including its energy source, processing units, and mechanical operations. When engineers are given examples are later asked whether the examples were helpful, most agree they were.[56] Yet a comparison of their designs to the designs of those who worked from scratch suggests the opposite.

Most of us will never design novel products, at least not in a professional capacity, but our knowledge of familiar products can lead to another kind of fixation, where we fixate on a product's intended function and overlook other potential functions. When we see a box, for instance, we think of it as a container, but a box can have many other functions, including drum, boat, seat, stage, table, cradle, trashcan, umbrella, costume, canvas, terrarium, flower pot, cat house, pedestal, projectile, or tunnel. These alternative functions are surprisingly difficult to imagine, even when the situation calls for one.

Suppose you need to affix a candle to the wall, and all you have are some matches and a box of tacks. What do you do? Tacking the candle to the wall is a nonstarter because the candle is too thick. Melting the candle to the wall is also a nonstarter because the candle is too heavy. The only feasible solution is to use the box, by emptying it of tacks and then tacking it to the wall. The candle can be placed inside the box, turning it from a container into a support.

This task, developed by psychologist Karl Duncker, is known as the candle problem.[57] It is administered in one of two ways: the way just described or a way that decreases the salience of the box's function as a container, by placing the tacks next to the box rather than inside it. The difference is subtle but substantial. Participants are twice as likely to solve the problem if the box is presented empty, and they solve the problem twice as fast. Seeing the box function as a container blocks our ability to imagine alternative functions—a disposition known as "functional fixedness."

Functional fixedness appears to be universal. People who live in technologically sparse societies, like the Shuar of the Amazon rainforest, have difficulty conceptualizing a box as a support even though they encounter boxes much less frequently than adults in industrialized societies.[58] Functional fixedness has also been observed in children, though, curiously, it's only observed in children aged six and older.[59] Six-year-olds are slower to use a box as a support if the box is presented full than if the box is presented empty. They're also slower to use a straw as a rod to push something through a tube, if the straw is presented in a cup than if presented by itself. Five-year-olds, on the other hand, are not impeded by observing a box used as a container or a straw used as a drinking device. They discern an alternative use for these objects as quickly as if the objects had been presented by themselves.

Are five-year-olds unaware of the intended functions of boxes and straws? Certainly not. By five, they would have used boxes and straws in their intended ways on countless occasions, and they would have observed others do the same. In fact, children as young as two know the intended functions of common tools, and they laugh (or protest) when those tools are used for other functions, such as when a toothbrush is used for painting or a key is used to stir food.[60] Rather, five-year-olds seem to have a more fluid conception of what a tool is for. Seeing a tool used for its intended function doesn't block their ability to devise alternative functions as it does for older children and adults.

This finding, on its face, seems to contradict the finding discussed earlier that young children have difficulty innovating. Five-year-olds' immunity to functional fixedness implies they conceive of tools more imaginatively than older children, contemplating not just what it is for but what it could be for. The catch, however, is that five-year-olds are no faster than six-year-olds to think of boxes as supports or straws as rods when these objects are presented by themselves. When the tool's intended function is *not* primed, children of all ages conceive of alternative functions quickly and accurately. So do adults.

Functional fixedness is really an error of commission, not omission. It arises from using a tool as intended or watching others use the tool as intended. Indeed, the longer we go without seeing a tool used for its intended function, the more successful we are at discovering alternative functions.[61] We are also more successful if our attention is directed away from the tool's intended function and toward its structure—its parts and materials, which can be coopted for other uses.[62]

Even superficial cues, like how a tool is labeled, can redirect attention from intended functions to alternative ones. In one study, participants were asked to complete an electric circuit using wire and a wrench.[63] The wire they were given was not long enough to complete the circuit, but they could complete it by using the wrench as a conduit because wrenches, like wire, are made of metal. Very few participants intuited this solution on their own. But when participants were asked to refer to the wire and the wrench by nonsense labels—"jod" and "beem"—they were significantly more likely to solve the problem. Calling a wrench a beem allowed them to bypass knowledge of its intended function and think of it as functionally similar to wire. A wrench by any other name may look the same, but it does not function the same, at least not in our imagination.

Innovation Accumulation

If humans are prone to copying unnecessary actions when learning about tools, prone to mimicking known examples when designing tools, and prone to fixating on known functions when using tools, how do we ever innovate? The answer lies not at the level of the individual but at the level of the group, through the cumulative effects of cultural evolution.[64] Individual humans rarely stumble upon innovative ideas, but when we do, others take notice, and the innovation spreads far and wide. In fact, most innovations do not emerge whole cloth but in a series of steps, as tool makers tinker with the designs inherited from their forebearers.[65] Over many generations, these incremental adjustments accumulate into large-scale innovations.

The modern sewing needle is a wonderful example of cumulative innovation.[66] This tool looks simple enough to have been invented in one afternoon, but it actually took hundreds of years and many incremental forms. The first sewing implements date back to Paleolithic times, over 25,000 years ago, and were made of bone, antler, and ivory. The crudest version consisted of two tools: an awl and a fork. The awl was used to poke holes in animal skins, and the fork was used to push thread through the holes. This design was improved by combining the awl and the fork into one device, with the awl on one end and the fork on the other.

A more significant improvement came from abandoning the fork and replacing it with a hole in the middle of the awl so the thread could be carried through the hole with each poke. This

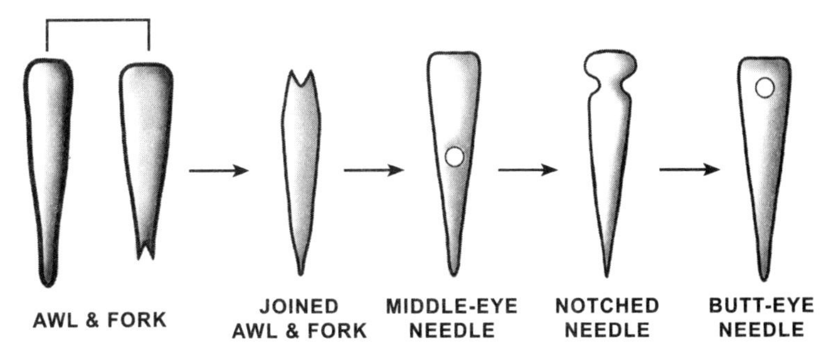

Figure 3.3 The modern sewing needle looks simple enough to have been invented in a single day by a single inventor, but it actually took hundreds of years and many incremental forms. (Adapted from Weber and Dixon 1989) (Courtesy of the author)

design proved brittle, however, and was abandoned for a version in which the hole was replaced with a notch at the top of the awl, which served the same purpose. But the notch made small sewing jobs onerous, as sewers had to knot their thread to the notch. The notch was thus replaced with a hole at the top of the awl, yielding the "butt-eye" design still in use today. This design allows for optimal hole making, needle threading, and thread pulling.

The invention of the sewing needle exemplifies the process of innovation in general. New technologies arise from successively tinkering with old ones. Some tinkering leads nowhere, but other tinkering yields improvements that accumulate and pave the way for additional improvements. Scholars of cultural evolution liken this process to a ratchet, or a mechanical gear that enables incremental motion in a desired direction.[67] Just as ratchets raise a load one notch at a time, cultural evolution raises our collective knowledge one innovation at a time.

Cultural ratcheting is uniquely human. Other animals may design fantastic structures, such as beaver dams, termite mounds, or spider webs, but these structures rarely vary across individuals or environments, suggesting that knowledge of how to build them is innate.[68] Natural selection invented the spider web, not the spider. An animal that comes close to cultural ratcheting is our closest living relative, the chimpanzee. Primatologists have observed marked differences in how chimpanzees from different regions of the world forage, groom, and court. They use tools and techniques that are widespread within their community but absent from other communities—they have their own cultures.[69]

Chimpanzees in captivity also develop their own cultures. They copy the foraging techniques modeled by other members of their colony, even when alternative techniques would be equally efficacious.[70] This copying behavior yields cultural differences between communities, but it won't yield a ratchet effect if community members don't tinker with, and elaborate on, the copied behaviors. And they don't. Chimpanzees can accrue one layer of innovation, but they can't accrue any more, potentially because they lack the uniquely human abilities of talking and teaching.[71]

Humans' knack for cultural ratcheting can be seen even in children.[72] Psychologists have studied this behavior with "diffusion chains," in which they teach a child a novel problem-solving technique and then observe whether other children copy this technique through a series of one-on-one interactions, similar to passing a message in the game of telephone. Messages tend to get garbled as they pass from person to person, but innovations do not. Preschoolers who are shown an effective technique for opening a puzzle box and retrieving a prize faithfully pass on that knowledge to their peers.[73] Other techniques are possible, but preschoolers copy the technique they see their peers using, from the beginning of the diffusion chain to the end.

Intriguingly, if preschoolers are taught a technique that includes unnecessary actions, they repeat the actions in the short term but not the long term, dropping them from the diffusion chain two to three links in.[74] Overimitation may thus be self-correcting, at least when solving culturally recurrent problems over several generations.

The reason that unnecessary actions drop out of the diffusion chain is that children do not just copy others' innovations but also tinker with them. They decompose the techniques they observe into smaller elements and then recombine those elements in novel ways. Children as young as four will recombine the elements of a multistep technique for opening puzzle boxes or building towers into something novel.[75] Children at this age are poor innovators on their own, but they will tinker with innovations modeled for them. They also tinker when working in pairs, building on the innovations of their partner. Indeed, five-year-olds faced with the hook task are twice as likely to succeed if they work with a peer.[76]

By copying and tinkering with others' innovations, children help propel the process of cultural ratcheting from one generation to the next. But sometimes the ratchet gets stuck. Old innovations, when faithfully imitated, preclude the need to discover new ones. For instance, children in diffusion chains do not always hit upon the best solution to a problem; sometimes they get stuck on a suboptimal solution in a kind of cross-generational Einstellung. Whether opening puzzle boxes, transporting rice, or building towers, children who observe suboptimal, yet adequate, techniques will happily copy them and pass them on to their peers.[77]

Imitation is thus a prerequisite for cultural innovation but also an obstacle. Humanity has transformed the world with mind-blowing technologies, developed through centuries of imitative tinkering, yet individual humans have difficulty understanding these technologies, let alone improving on them. Our technological imagination may be broader now than ever before, but our propensity to imitate limits whether, and how, we apply that imagination to discovering further innovations.

4 Anomalies

Expanding Our Empirical Imagination

In October 2017, astronomers in Hawaii observed an unusual object pass through our solar system.[1] They called it ʻOumuamua, the Hawaiian word for scout. ʻOumuamua was assumed to be an asteroid ejected from another solar system, but it was longer, shinier, and slower than an asteroid should be. ʻOumuamua was ten times as long as it was wide and ten times shinier than a typical space rock. It was also traveling much slower than it should have if ejected by a random collision. Even more unusual, ʻOumuamua was observed to change paths. Comets sometimes change paths as they expel gas, but ʻOumuamua showed no sign of a gaseous tail.

If ʻOumuamua is neither an asteroid nor a comet, what could it be? Harvard astrophysicist Avi Loeb has put forward a radical hypothesis: ʻOumuamua is a piece of alien technology, specifically, a metal sail propelled by solar radiation that was launched into interstellar space by an advanced alien civilization. If there are other intelligent beings in the universe, as most astronomers believe there are, then it's conceivable that some of those beings have developed technology for exploring outer space, as we have.

Loeb's hypothesis has been greeted with skepticism by his peers, who think that 'Oumuamua's behavior can be explained without appealing to any extraterrestrial intelligence.[2] Loeb and his colleagues have published analyses suggesting that the likelihood of 'Oumuamua being a naturally occurring object is less than one in a trillion, but many astronomers would prefer to accept those odds than the possibility that we have stumbled upon alien technology—or, more aptly, that alien technology has stumbled upon us. Still, those who reject the possibility that 'Oumuamua was crafted by aliens must grapple with how to explain its unusual behavior, which also requires a stretch of the imagination.

Or we could ignore 'Oumuamua. Discoveries that lie outside the boundaries of what we know to be possible are more easily dismissed than embraced. We have strong expectations about how the events around us should unfold, and when we observe them to unfold differently, we often privilege expectation over observation. Revising our expectations takes work whereas ignoring an observation takes none. Anomalies can expand the boundaries of what we deem possible but only at the expense of the expectations defining those boundaries.

Historically, scientists have been slow to embrace anomalies that would come to change their field.[3] These anomalies include Lavoisier's discovery that oxygen is required for combustion, which revolutionized chemistry; Roentgen's discovery that electric currents can produce X-rays, which revolutionized physics; and Mendel's discovery of recessive traits, which revolutionized biology. Anomalies reveal to scientists that the space of possibilities within their field is larger than their current theories permit. But they don't dictate how to account for the new possibilities, and there are many methods of accounting that leave one's theories intact.

Consider the resistance of early twentieth-century geologists to the discovery that fossils of the same species can be found in distant lands—lands now separated by ocean.[4] Cambrian trilobites, Permian ferns, and Triassic lizards appear to have existed in both Europe and North America or in both Africa and South America. Biologists even discovered the same living species on opposite sides of an ocean, including beetles, earthworms, and opossums. One explanation for this dispersion of lifeforms is that the earth's continents have not always been separated by ocean but were once part of a single landmass (Pangea). But this explanation was too incredible without an accompanying explanation for how the continents had moved. So geologists explained the dispersion in a different way: land bridges, from one continent to another. These bridges would have to have been stable enough to span 3,000 miles of ocean but not stable enough to survive to modern times.

The natural historian Stephen Jay Gould notes that "the only common property shared by all these land bridges was their utterly hypothetical status; not an iota of direct evidence supported any one of them."[5] But he also notes that, to a geologist of the 1930s, the one thing more absurd than imaginary land bridges was the idea that the continents had drifted apart. "*Impossible* is defined by our theories," writes Gould, "not given by nature."

Today, land bridges are the impossibility and continental drift is the theory that precludes them, but that change took decades of rethinking and reanalysis, not to mention tons of new data from satellite observations of the earth's crust and oceanographic surveys of the seafloor. The path from anomaly to orthodoxy is a long one, pithily summarized by psychologist Chaz Firestone as "(1) that's false, (2) that's trivial, (3) I thought of it first."

Scientific anomalies take months, years, or even decades to be appreciated because scientists are inherently cautious in changing their views, as we all are. This caution can arise from an unwillingness to accept that we might be wrong, but it can also arise from legitimate skepticism about the anomaly. If dozens of considerations imply that the earth's continents are fixed in place, why should geologists disregard them all in light of one new discovery? The discovery could be flawed, not the orthodoxy it challenges.

Consider the "discovery" of N-rays, a fictional form of radiation thought to be analogous to X-rays.[6] N-rays were documented and described in several dozen scientific publications in the early 1900s before they were revealed to be an artifact of biased observation. The discoverer of N-rays was physicist Rene Blondot from the University of Nancy, after which the rays were named. While experimenting with X-rays, Blondot thought he had observed a form of radiation that could be bent with a prism. This radiation appeared to be emitted by a variety of sources, including white light, heated silver, and the human body. Blondot performed experiments on N-rays by manipulating these sources and observing any changes in radiation registered by nearby photodetectors.

The problem with these experiments is that the photodetectors were checked by eye, and researchers observed changes only under the right conditions and with the right training. A skeptical visitor to Blondot's lab, the physicist R. W. Wood, was unable to verify these observations himself. He then went on to demonstrate that the observations were illusory by secretly removing a critical part of the experimental apparatus—the prism that refracts N-rays—and watching to see whether his ruse influenced what the researchers saw. It did not. A whole community of physicists had unwittingly convinced themselves that they could see a ray that does not exist.

A more recent example of a false discovery is the 2014 article "When Contact Changes Minds: An Experiment on Transmission of Support for Gay Equality," published in the preeminent journal *Science*.[7] The researchers described a study in which people's attitudes toward gay marriage became more favorable after a single conversation with a gay person advocating for gay marriage. These findings were covered by several media outlets, including the radio show *This American Life*.[8]

The host of *This American Life,* Ira Glass, introduced the study by flagging it as an anomaly. "When it comes to the big hot button issues," said Glass, "do you know anybody who has changed their minds? . . . Probably not, right? In fact, the opposite happens. There's this thing called the backfire effect. It's been documented in all kinds of studies. It shows that when we're confronted with evidence disproving what we believe, generally we dig in and we believe it more. And the rare times that people do change, it's slow. You don't just have an argument with your uncle over the invasion of Iraq, and then at the end of dinner, one of you goes, 'OK, I no longer believe what I did; you're right.' People just don't flip like that, which is why this [study] is so incredible."

Glass went on to describe the study, noting that voters not only changed their position while being interviewed but also retained that position a year later. They even managed to convince other members of their household to switch positions. "Apparently neither of those things ever happens," he explained.

I listened to this show while driving in the car with my family, and they were shocked when I began yelling at the radio. "If neither of those things ever happens," I yelled, "then why are you covering this study on your show?!" As a social scientist, I was skeptical that people's position on a divisive issue could be enduringly changed with a twenty-minute conversation, and it seemed irresponsible to broadcast this finding before it was replicated by other researchers, let alone reconciled with dozens of other studies that had failed to achieve similar outcomes.[9]

Only a month later, it came to light that the data reported in the *Science* publication had been made up. No one's attitudes had actually been measured, let alone changed. I felt vindicated in my initial rejection of the study but also aware of the irony that I refused to change my mind about a study purporting to change minds. What evidence *would* I find convincing?

I was right in this particular case to stick with my convictions, but I might not have been. A gut reaction that an anomaly is wrong is not evidence, in and of itself. Oxygen, X-rays, recessive traits, and tectonic plates were all met with disbelief at the time of their discovery, but none turned out to be wrong. How we reconcile anomalies with the expectations that make them anomalous is neither obvious nor easy, but the process has the potential to expand imagination, should we engage with it.

Anomaly Animosity

It's not just scientists who have a complicated relationship with anomalies. The psychologists Clark Chinn and William Brewer identified eight ways the average person deals with anomalies, and only one involves changing our beliefs.[10] The other seven are ignoring the anomaly, rejecting it, excluding it, professing uncertainty about it, holding it in abeyance, reinterpreting it, or accommodating it peripherally by changing collateral beliefs. Throw an anomaly at us, and we'll bat it off with speed and precision.

What would you say, for instance, to the possibility that dinosaurs were warm-blooded? Dinosaurs were anatomically similar to modern-day reptiles, which are cold-blooded, but an analysis of the density of dinosaur bones suggests that dinosaurs grew faster than reptiles, at a rate more similar to warm-blooded mammals. Students who read about this anomaly in one of Chinn and Brewer's studies had many ways of discounting it and thus defending their conviction that dinosaurs were cold-blooded, including

- "There is too much evidence that concludes to dinosaurs being cold-blooded" (ignoring the data).
- "I don't feel that there is a direct correlation between being warm-blooded and growing quickly" (rejecting the data).

- "These are only two kinds of dinosaurs, and they don't represent the whole category of dinosaurs" (excluding the data).
- "This can be true only if we're sure bone density is always parallel with growth rate" (professing uncertainty about the data).
- "There might be other explanations as to why dinosaurs' bones are less dense other than the fact that they were warm-blooded" (holding the data in abeyance).
- "The bones have been fossilized, which might alter the current state of structure of the bone tissue" (reinterpreting the data).
- "Evolution could have changed bone density and growth rates; cold-blooded animals may have had a fast growth rate at one point" (peripherally accommodating the data).

Each of these strategies for discounting anomalies is, in fact, a collection of strategies. If we are motivated to reject the anomaly, we might pick at the quality of the sample or sampling methods, question the accuracy of the measures or measuring assumptions, or raise objections about the researchers' expertise or motives. If we are willing to accept the anomaly but unwilling to revise our theory, we could reinterpret the anomaly by appealing to alternative causes or auxiliary hypotheses, question the anomaly's scope or generalizability, or claim the anomaly is actually consistent with our theory or, better yet, predicted by it.

We deploy these strategies even as children.[11] Children who believe that objects sink or float depending on their weight will dismiss the observation that heavy objects sometimes float, claiming these objects must weigh less than they thought.[12] Children who believe that blocks always balance at their geometric center will dismiss the observation that irregularly shaped blocks balance elsewhere, claiming the blocks must have magnets inside.[13]

A child who appeals to magnets as a way of preserving erroneous beliefs about gravity strikes us as stubborn, even comically so. But privileging prior beliefs over contradictory evidence can be a rational course of action, as noted earlier. One of my favorite illustrations as to why comes from the children's sitcom *Jessie*. A character on the show is nicknamed Crazy Connie for her obsessive behavior toward her classmate Luke—behavior that includes lying, stealing, trespassing, and vandalizing. In one episode, Connie tries to convince everyone she's turned a new leaf and is no longer obsessed with Luke. Later, however, they discover she's kidnapped him. "I knew Crazy Connie was still crazy!" exclaimed Jessie, the show's protagonist. "My first clue was everything she's ever done."

If the world tells you, with every observation, that your theory is correct, then it is rational to discount the rare observation that does not accord with that theory. Still, the fact that we have seven different strategies for discounting anomalies suggests that we are not always acting rationally. One behavior that is particularly difficult to justify on rational grounds is that we often fail to register anomalies *as* anomalies. We fail to see the inconsistency between what we see and what we believe.[14]

In one study of this behavior, physics students answered physics questions before and after watching a video tutorial on force and motion.[15] The questions targeted common misconceptions about the relations between force, velocity, and acceleration, such as this one: "Consider a juggler tossing a ball in the air. After the ball leaves his hand, the force on the ball is (a) upwards and constant, (b) upwards and decreasing, (c) downwards and constant, or (d) downwards and decreasing." The correct answer is (c), "downwards and constant," because the only force acting on the ball is gravity, but many participants selected (b), "upwards and decreasing," on the assumption that the ball has acquired a force—"the force of motion"—that keeps it moving upward after it has left the juggler's hand but only for a limited time.

After answering a series of such questions, the participants watched a video explaining the force dynamics of projectile motion, in which it was explicitly stated that, aside from air resistance, "only one force acts on a ball throughout its flight: the force of gravity, which is constant and downwards, accelerating the ball in the downward direction." The participants described the video as "clear," "concise," and "easy to understand," and their confidence in their answers increased from pretest to posttest. But the participants continued to provide incorrect answers at posttest, exhibiting no improvements in understanding.

When the participants were asked to explain their answers, they seemed oblivious to the inconsistency between the video tutorial and their prior beliefs. According to one participant, "The video . . . said the ball is slowly decreasing in force so therefore it stops at one point and comes down." Another participant noted that "it wasn't that hard to pay attention to [the video] because I already knew what she was talking about."

Casual observation of contradictory information appears to be insufficient for people to register the contradiction. This finding has been documented consistently in the science education literature. It holds for people of various ages and educational backgrounds and for interventions targeting a wide range of misconceptions, including misconceptions about gravity, buoyancy, optics, and illness.[16] The challenge of reconciling prior expectations with unexpected findings is rarely undertaken, at the least when learning science.

Another irrational aspect of how we balance theories and observations is that we perceive evidence for our theories even when no such evidence exists. Our inability to see findings we do not expect to see is complemented by an ability to see findings we expect to see but are not real. Imagine, for instance, you are a school nurse treating students complaining of stomach cramps. The students arrive at your office soon after lunch, and you suspect that something they ate gave them food poisoning. You query these students and discover that all of them ate the school's tacos. You then report the incident to the school administrators, labeling the tacos as a health hazard. Are you justified in your conclusion?

Many people would find this evidence convincing, but there are other observations you need to make before the tacos can be identified as the cause of the students' stomach cramps. You need to know, for instance, how many students ate tacos but did not develop cramps. The handful of students who came to your office may be a small portion of all taco eaters, with the rest feeling fine. You also need to know whether any students who did not eat tacos developed cramps. If such students exist, then the cramps are likely due to some other shared experience, such as a strenuous gym class after lunch.

The belief that tacos caused the students' cramps seems more tenuous when we consider the full range of data needed to draw that conclusion, but the belief itself can blind us to the limitations of the data we do have. We are apt to read correlations into uncorrelated data even when the data are fully displayed before our eyes.[17] For example, in one study, people saw data indicating that the health of a plant depended on a charm but, ironically, did not depend on the sun; the plant always died if the charm was taken away but died only occasionally if removed from the sun. Still, people judged the sun to be more important to the plant's health than the charm, especially children.[18] Eight-year-olds were so convinced that the sun mattered and the charm did not that their judgments of what caused the plant to stay healthy remained unchanged regardless of what data they saw.

The effect of expectations on data evaluation can be seen at a neurological level as well, in how the brain processes expected and unexpected data. When we encounter expected data, such as the observation that antidepressants improve one's mood, we recruit neural circuits in the parahippocampal gyrus. But when we encounter unexpected data, such as the observation that antibiotics improve one's mood, we recruit a different set of circuits—those in the anterior cingulate, left dorsolateral prefrontal cortex, and precuneus.[19] If we interpreted data as mere information, then one might expect to find similar patterns of activation in both cases. But the finding of different patterns suggests that we process data in light of our expectations, as confirmation or disconfirmation thereof.

Expectations thus influence what we see and how we see it. They can also influence how we behave, steering us away from seeing anything unexpected in the first place. Decades of research on classroom instruction have revealed that when students are left to explore a topic on their own they rarely discover anything unexpected. Instead, they revisit findings they've already observed and apply principles they've already mastered.[20] Although many instructors are enamored with the idea of students discovering hidden truths for themselves, the pedagogical technique of "discovery learning" does not work; it yields neither discovery nor learning.

To see why, consider the performance of Joey, a middle schooler in a discovery-learning study tasked with determining what factors influence how quickly a boat moves through a canal.[21] Joey was given a canal with an adjustable bottom and boats of different sizes, weights, and shapes and shown how to time the boats as they sailed through the canal on a pulley. A systematic approach to investigating boat speed would be to vary one factor at a time and compare speeds for just that variation, but Joey varied multiple factors on each trial. His first trial consisted of a small, circular, light boat in a deep canal and his second consisted of a large, square, heavy boat in a shallow canal. From these trials, he inferred that the second boat would have gone faster if it were lighter, a hypothetical outcome he had not observed nor would go on to test.

Because Joey varied multiple factors at once, the inferences he drew from contiguous trials were always invalid. And many of his trials duplicated previous trials because his experiments were unplanned. Joey's lack of systematicity was not a fatal flaw; he could still stumble upon observations that might challenge his prior beliefs. But Joey proved inattentive to such observations. He began the experiment convinced that the only factor that mattered was boat size, and he ended with the same conviction. Joey spent most of his time demonstrating, to his own satisfaction, that small boats move faster than large boats, ignoring confounds in his experiments and evidence (from noncontiguous trials) that other factors mattered, too.

Joey's failure to learn from self-directed exploration is typical for this type of instruction. A meta-analysis of 164 discovery-learning studies found that "unassisted discovery does not benefit learners."[22] Students learned substantially more from direct teaching, such as lectures and demonstrations, than they learned from self-directed exploration, à la Joey.

That said, the most successful kind of instruction was not direct teaching but a combination of teaching and exploration—a technique coded as "enhanced discovery." The success of enhanced discovery puts a finer point on what's wrong with unassisted discovery. Enhanced discovery, by definition, included some form of external guidance, whether it be prompts to explain, prompts to reflect, worked examples, progress checks, or regular feedback. These scaffolds constrain students' exploration, ensuring they will discover ideas that challenge their expectations and, accordingly, expand their imagination. Whereas unassisted discovery allows students to fixate on what they already know, enhanced discovery forces them to engage with something new. It's the difference between taking a guided tour of a foreign land and wandering aimlessly in your own neighborhood. We will return to this difference in chapter 8 from the perspective of whether free play facilitates learning (spoiler alert: it doesn't).

One-Track Minds

Anomalies can reshape our map of empirical possibility, just as tools can reshape our map of technological possibility and testimony can reshape our map of historical possibility. The challenge in all of these cases is figuring out how to connect the new with the old, the unexpected with the expected. This process is complicated by our disposition to discount unexpected findings—by ignoring them, rejecting them, failing to see them, or failing to seek them out—but it's also complicated by the inherent difficulty of coordinating observations and explanations.[23] Even if we're willing to accept an anomaly, it's almost never obvious how to account for it.

There are three challenges in coordinating observations with explanations. One is figuring out how to use observations to navigate the space of all possible explanations. If, for instance, you are trying to determine why some boats sail through a canal faster than others and you observe that light boats sail as fast as heavy boats (all else being equal), then you should abandon the explanation that weight makes a difference and move to other possibilities, such as boat shape or boat size. Another challenge is figuring out how to use explanations to navigate the space of all possible observations. If you decide that boat weight doesn't matter but suspect that boat shape does, then you should stop manipulating boat weight and start manipulating boat shape, comparing, for instance, round boats to pointy boats while holding other factors constant.

These two challenges assume you've resolved an even more fundamental challenge: learning to differentiate observations *from* explanations. Joey, from the earlier study, failed to discover that boat speed is influenced by factors other than boat size because he did not explore the situation in a way that would allow him to differentiate observations from explanations. Controlling the situation, by manipulating one variable at a time, is crucial to that endeavor, but so is using what you see to evaluate what you believe and using what you believe to guide what you see. Joey showed no sign of coordinating these distinct spaces of information; instead, he bounced between them, explaining some observations but not others and testing (or attempting to confirm) some explanations but not others.

Disentangling observations from explanations is difficult for adults as well as children. When asked to justify our beliefs, we typically provide an explanation rather than cite evidence.[24] Suppose you believe that small classes lead to better learning. If I asked you why, you might tell me that students get more attention in small classes, that they are more likely to participate in class discussion, or that they are less likely to get distracted by their peers. But you probably wouldn't describe a study demonstrating the link between class size and learning outcomes.

Granted, you probably do not know of such a study, but if I pressed you on the issue, asking what evidence might support your belief or what evidence you would want to convince others of your belief, you would probably still cling to your initial explanations, elaborating on the mechanisms behind them rather than describing a relevant source of data. It doesn't matter that teachers in small classrooms can provide more tailored instruction and more feedback on assignments if, it turns out, there is no correlation between class size and student learning.

Our focus on explanations over observations makes sense from a practical point of view because explanations transcend observations; they provide a means of organizing and unifying our observations, rendering them applicable to new situations. Without explanations, we would be forced to compare events one by one, and we would rarely glean the commonalities among them. But how can we keep the explanations we prefer from interfering with the observations we might make? The answer is not to forget our preferred explanations, if that were even possible, but to seek out more explanations—explanations grounded in different considerations, allowing for different interpretations of the observations at hand.

Alternative explanations turn out to be key to accepting the unexpected.[25] They provide a ready-made path from unexpected observations back to expected ones within the landscape of all possible observations. Studies have found that alternative explanations allow us to interpret observations that would otherwise be uninterpretable, leading to greater confidence in the observation as well as the explanation. They also lead us to reassess our original beliefs, reducing confidence in those that do not align with the newly explained observation. The ease with which we use alternative explanations to recalibrate our expectations is the reason why magicians never reveal their secrets. The extraordinary quickly become ordinary when explained.

Even children appreciate the value of alternative explanations.[26] Five-year-olds, who believe that objects sink or float depending on their weight, will not change their mind after watching objects of the same weight sink or float depending on their shape. But if those same children learn about buoyant forces and how larger objects are more susceptible to buoyant forces, then they are likely to change their mind that weight alone determines buoyancy and begin to take shape into account. Critically, children change their mind only if they learn the alternative explanation before they witness the unexpected events. If they witness those events first, they are rarely able to reinterpret them later from memory.[27]

Coordinating explanations and observations thus involves one further challenge: coordinating alternative versions of each. Just as we have to coordinate observations with explanations and explanations with observations, we also have to coordinate multiple observations and multiple explanations. This challenge requires practice and attention.[28] It also requires recognizing the alternatives *as* alternatives—to mark each possibility as exclusive and distinct.

As much as children benefit from learning about alternatives, they have trouble keeping them distinct in their heads. They tend

Figure 4.1 Four-year-olds recognize that a ball dropped through a forked tube could exit on the left or the right and put their hands under both openings, but younger children put their hands under just one opening. (Adapted from Redshaw and Suddendorf 2016) (Courtesy of the author)

to fixate on one possibility and ignore the relevant alternatives, especially when they are young. Take, for example, the simple task of catching a ball dropped through a forked tube.[29] In one set of studies, two-year-olds were introduced to a tube shaped like an inverted Y and shown that when you drop a ball into the opening at the top, it can come out either opening at the bottom. The toddlers were then asked to place their hands under the tube to catch the ball before it hit the ground. The optimal placement is to put one hand under each opening, but they put both hands under a single opening, fixating on one possibility and ignoring the other.

These studies included twelve trials, so the toddlers who failed to cover both openings could watch the ball fall through the uncovered opening and learn from their mistake. Some toddlers do learn from their mistake and begin to cover both openings, but the majority continue to cover just one. And some who begin to cover both openings regress back to covering just one in later trials. Their performance mirrors the performance of chimpanzees, who also cover a single opening and rarely learn to cover both. Four-year-olds, on the other hand, typically cover both openings from the start, and those who do not almost always learn to cover both after a few trials. Four-year-olds are thus able to juggle multiple possibilities in a way that toddlers and chimpanzees are not.

You might be concerned that toddlers realize the ball could fall through either opening but think they need to use both hands to catch it. Physically, they should have no problem: the ball is the size of a grape (or is an actual grape, when chimpanzees are tested), and the opening of the tube is small enough for toddlers to cover with their palm. Still, researchers have addressed this concern by comparing performance on the forked-tube task to performance on a double-tube task, where two balls are dropped into two adjacent tubes on each trial. The double-tube eliminates the uncertainty inherent in the forked-tube, and toddlers respond immediately and consistently. Half put both hands under both openings on the first trial, and all of them learn to do so by the end of the session. The double-tube task turns two possibilities (left opening *or* right opening) into one (left opening *and* right opening), and they succeed.

Other studies, involving very different task demands, have documented similar failures in children's ability to keep track of multiple possibilities, even among older children.[30] When three-year-olds are shown a forked slide and are asked to put cotton at the end of the slide to break the fall of a toy mouse, few put cotton under both sides of the fork.[31] When four-year-olds are shown a picture drawn in red marker and are asked which of three children drew it—a child holding a red marker, a child holding a blue marker, or a child holding a marker whose color cannot be discerned—most claim it was the child with red marker and insist they know for sure, even though the task was constructed so they couldn't know for sure.[32]

To be clear, children have no problem entertaining multiple possibilities; they easily recognize that a ball could fall through either opening of a forked tube or that a mouse could travel down either side of a forked slide. But they have problems coordinating these possibilities when reasoning about them and responding to them. They fixate on one possibility and treat it as certain rather than mark all possibilities as viable.[33] Possibilities do not become truly possible until we think about them in the context of their alternatives.

The Feeling of Impossibility

For all the problems we face in acknowledging and accepting unexpected events, we do have one psychological mechanism working in our favor: awe. This emotion is triggered by the violation of a deeply held expectation and motivates the search for information that might repair the violation. It leads us to embrace an anomaly rather than reject it, and to undertake the difficult task of coordinating observation and explanation as we explore its source. Many scientists report being driven by awe.[34] An awe of combustion led Lavoisier to rethink the structure of matter. An awe of electromagnetism led Faraday to rethink the properties of energy. An awe of biodiversity led Darwin to rethink the origin of species.

Awe is a complex emotion built upon two other emotions: surprise and curiosity. When the world delivers an unexpected outcome,

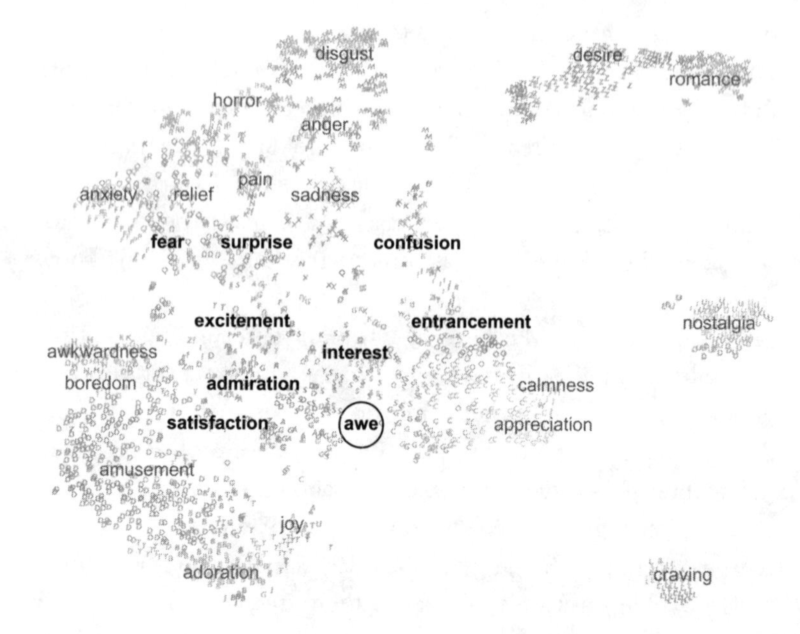

Figure 4.2 Research mapping people's emotional reactions to brief events shows that awe is related not only to positive emotions, like admiration and excitement, but also negative emotions, like confusion and fear. (Adapted from Cowen and Keltner 2017) (Courtesy of the author)

we feel surprise; and if we are unable to explain the outcome, we become curious. The bigger the discrepancy between what we expect to observe and what we actually observe, the bigger the surprise.[35] And the more difficulty we experience in resolving that discrepancy, the more curious we become (up to a point).[36] Discovering that your dog ate a hamburger off the counter might elicit surprise, but it wouldn't elicit curiosity. You know who ate the burger and why. Discovering that your dog *put* a burger on the counter would elicit both surprise and curiosity. Where did he find the burger? Why did he put it on the counter?

Curiosity turns out to have wide-ranging benefits for thinking and learning.[37] It enhances attention, promotes exploration, prompts explanation, and increases memory.[38] If there's an answer to be found, curiosity will help us find it. Consider what might happen if you saw your dog put a burger on the counter. You would stop what you're doing and pay full attention to the dog. You'd follow him around and look for the source of the burger. All the while, you'd be pondering explanations for the dog's behavior: maybe he found the burger in the trash; maybe he grabbed the burger off the ground on his last walk. When you finally figure out what happened—that the burger is, say, a realistic toy you forgot you owned—you will remember the event better than other events that day or even that week.

The benefits of awe can be observed even in infancy.[39] In one study, eleven-month-old infants watched events that appeared to violate the laws of physics and were then allowed to interact with the objects involved in those events. They saw, for instance, a toy train that appeared to pass through a solid wall and were then allowed to play with either the train or a new toy. They not only preferred to play with the train but also played with it in a way that tested its solidity, such as banging it on the ground. When they saw a train appear to float in the air, they played with it in a way that tested its gravity, by dropping it and watching it fall. In addition to exploring expectation-violating toys, infants scrutinized the properties of these toys, learning, for instance, whether the toy played music. They showed no such interest in toys that conformed to their expectations.

Curiosity evokes exploration and explanation because we find it intrinsically rewarding to resolve the surprise that triggered it. When researchers have explored curiosity's effect on the brain, they've found that curiosity activates areas associated with external rewards, like food and money.[40] Seeking information in the wake of curiosity is so rewarding that we will do so even if it hurts us. We are eager to know how much money we lost by not taking a bet, even though knowing can only lead to regret, and we will willingly receive an electric shock to find out whether the pen in front of us is a normal pen or a prank electronic-shock pen.[41]

If curiosity is such a powerful drive—powerful enough to kill the cat, as the saying goes—then why isn't curiosity our typical reaction to unexpected information? Why are we so likely to distrust that information or disregard it? The answer is complicated.[42] It depends on how much we anticipate learning from the anomaly, how useful that knowledge might be in the future, and whether that knowledge conforms to our prior beliefs. The sinking of the RMS *Titanic* is the kind of anomaly that reliably evokes curiosity. We think there is much to learn, that this learning could be useful in the future, and that it will accord with our prior beliefs about boats, icebergs, and human calamities. How a ship as massive as the *Titanic* was able to float in the first place is also a mystery, given our naive belief that heavy objects sink, but this mystery evokes less curiosity. We expect the answer would be difficult to learn, difficult to apply beyond the context of shipbuilding, and difficult to reconcile with our prior beliefs.

Still, if we go to a port and watch ships sailing on the ocean with our own eyes, we may well become curious through the awe evoked by the ships' massive size. Awe, like curiosity, is triggered by surprise, including the surprise of perceiving something immense or vast. But awe transcends curiosity in its sense of mystery. We feel awe when we recognize not just that there are gaps in our knowledge but that we have no idea how to fill those gaps.[43] How could something as large as a skyscraper glide through water as gracefully as a swan? What aspect of a ship's construction offsets all that weight? Our naive understanding of buoyancy cannot accommodate the physics of a ship, and awe is the experience we feel when we recognize that inconsistency—when we recognize we've observed something that our theories of the world deem impossible.

This feeling of impossibility drives us to do more than just acquire information—it drives us to change our understanding.[44] Awe inspires us in a way that curiosity does not. Curiosity comes and goes whereas awe can sustain a lifetime of learning and discovery.

Some people are more disposed to feel awe than others, and those people tend to have a greater appreciation of science and a keener sense of logic.[45] But anyone can experience awe if the conditions are right. Good science educators and science communicators know how to trigger awe in their audience by honing the expectations violated by an awe-inducing event.[46] All of us hold misconceptions about how the world works, so all of us have the potential to experience awe when those misconceptions are laid bare. Even commonplace observations have the power to induce awe if framed appropriately. We observe each day that the sky is blue, for instance, but we rarely stop to ponder why. How could the sky be blue if air is transparent? Why doesn't the air in a room appear blue? Why does the sky sometimes appear yellow or red?

With awe, unexpected discoveries can be used to expand imagination rather than close it. Awe turns an anomaly we would rather ignore into a mystery we have to solve. But this emotion has its drawbacks. We typically experience awe as a positive emotion, but it can also yield confusion, intimidation, and fear.[47] The line between awesome and awful can be thin, as when we are in the presence of an "awing power" or witness a display of "shock and awe." Grand religious spaces like mosques and cathedrals were built to inspire awe, but the kind that evokes reverence rather than inquiry and deference rather than discovery. Even when awe does inspire inquiry and discovery, it is fueled by an underlying sense of uncertainty, which many people find aversive. Accepting an anomaly's invitation to expand our imagination is not for the faint of heart. Doing so can itself be an act of awe.

Part II

Expanding Imagination by Principle

5 Science

Expanding Our Causal Imagination

Male widowbirds are an animal that stretches imagination. They have long black tails that hang twenty inches below their bodies as they fly over the African grasslands. Their tails make them unusually beautiful but also unusually vulnerable to predators like hawks and eagles. Their tails also make foraging cumbersome and impose metabolic costs that make the birds vulnerable to parasites.[1] Long tails doom male widowbirds to shorter lives, making widows of their mates (hence, their name).

Why would male widowbirds inherit a trait that dooms them to an early death? This question is difficult to answer regardless of whether you subscribe to creationism or evolution. From a creationist perspective, it would appear that God created widowbirds to be easy prey. Why would God be so cruel? From an evolutionary perspective, it would appear that widowbirds evolved a trait that makes them patently less likely to survive and reproduce. What selection pressure could give rise to such a trait?

The mystery of the widowbird cannot be resolved using the standard principles of intentional design or natural selection but instead requires a new principle: sexual selection.[2] This principle was first articulated by Charles Darwin in 1871 in *The Descent of Man* as an addendum to his principle of natural selection. It stipulates that mate preferences can exert pressure on the evolution of a species independent of selection pressures. If female widowbirds prefer long tails then males with longer tails will attract more mates and have more offspring than males with shorter tails. Longer tails might then confer a reproductive advantage even if they confer no survival advantage. At some point in evolution, widowbirds became embroiled in a tradeoff between tails long enough to attract a mate but not so long as to attract a predator. The longer a tail a male could sport without getting caught by predators, the more likely he caught a female.

This idea inspired experiments in which researchers clipped or lengthened a widowbird's tail and then tracked the number of offspring they produced.[3] Sure enough, widowbirds with clipped tails produced fewer offspring than those with normal tails, and those with normal tails produced fewer offspring than those with artificially long tails.

Sexual selection thus solves the mystery of the widowbird. But it's not the only mystery this principle solves. Many other traits that hinder an organism's survival turn out to bolster their reproductive success by attracting mates. These traits include the conspicuously loud calls of male frogs, the conspicuously bright plumage of male birds, and the excessively heavy antlers of male deer.

In the landscape of biological possibilities, widowbirds are a datapoint far from the possibilities we regularly observe and readily understand. They are an anomaly. Connecting this anomaly to what is familiar could be done in a way that's specific to widowbirds—perhaps the widowbirds' tail helps with heat regulation or aerodynamic control—but connecting it through the principle of sexual selection paves the way for many other connections. Widowbirds are no longer an isolated anomaly; they are one of many anomalies succinctly explained by the same principle. Sexual selection expands the landscape of possible traits in a systematic fashion, incorporating traits that previously defied imagination while also spurring the discovery of new traits.

Consider the unique trait of human speech.[4] Theories abound for why humans have language, but they take for granted the more fundamental question of why humans can make so many sounds—dozens more than other mammals. Our vocal tract is elongated relative to theirs, but what drove the elongation? Evolution could not have foreseen the benefits of language. It must have favored elongated vocal tracts for some other reason, and sexual selection provides a potential answer: longer vocal tracts produce deeper sounds and deeper sounds are indicative of larger bodies. If female humans preferred larger males to smaller ones and, accordingly, deeper voices to higher ones, then this preference could have driven the elongation of vocal tracts in general.

This account of the origin of human speech is speculative and may ultimately turn out to be wrong, but it illustrates the power of sexual selection to generate novel accounts of biological traits, and, more generally, the power of principles to generate novel ideas. Anomalies reveal that our map of possibilities is incomplete, and principles provide a means of completing it.

Principles can be misapplied, of course. Not all sex differences are the result of sexual selection, and sexual selection can lead us to posit sex differences that do not actually exist. The ideas generated by principles need to be verified. Still, the generative power of principles renders them superior to examples. If an example is a new datapoint, a principle is a new function, and the more functions we know, the more broadly—and more purposefully—we can apply our causal imagination.

Causes, Causes Everywhere

Not all principles are equally generative. A principle can be too broad or too narrow, too complex or too shallow. The principles we prefer are those that cover many things rather than just a few (abstract principles); those that confer a sense of understanding rather than just prediction (causal principles); and those that provide a sense of structure rather than just association (mechanistic principles). Each of these features—abstraction, causation, and mechanism—satisfies different criteria for what makes a principle useful. Abstraction allows us to transfer the principle from one instance to another; causation allows us to explain and control those instances; and mechanism allows us to connect and organize the instances across diverse contexts. We will explore each of these features in turn, beginning with causation.

Causation is a relation among events, where one event brings about another. Such relations help us predict future events, explain past events, and intervene on present events to change their outcome. Causal knowledge has obvious benefits—benefits even young children seek to obtain. As soon as children begin to utter complete sentences, around a quarter of those sentences contain causal information, either in the form of a question, such as "Why one piece broke?" or an explanation, such as "I got this candy because it's a prize."[5] When consulting other people for information, children consult those who have a record of providing causal explanations and avoid those who provide mere descriptions.[6] They actively evaluate the quality of the information they receive, asking follow-up questions if that information is vague, tangential, or circular.[7] And when they receive the causal explanations they were looking for, they commit those explanation to memory, forgetting any noncausal information learned along the way.[8]

Children's drive to learn causal information expands their repertoire of cause–effect relations and thus the scope of events they can understand and anticipate. But only some cause-effect relations can be grasped early in life. Children easily grasp relations involving contact causality, or the transfer of momentum from one physical object to another, and intentional causality, or the goal-directed activities of sentient beings, but they lack knowledge of other domain-specific causal relations, such as those governing life and death, reproduction and inheritance, velocity and acceleration, or heat and temperature.[9] We have a seemingly innate appreciation of the physical properties of objects and the intentional properties of agents, but we must learn the causal principles that underlie everything else.

In my book *Scienceblind,* I discuss the difficulty of acquiring causal principles that transcend those we know early in life.[10] Early acquired principles impede the learning of later acquired ones because they provide an alternative means of understanding the same phenomena. For instance, through the lens of intentional agency, we understand life as the capacity for goal-directed motion (rather than reproduction and metabolism) and lifeforms as products of intentional design (rather than evolution). Through the lens of contact causality, we understand objects as holistic units (rather than collections of molecules) and gravity as something that pulls objects down (rather than together).

Learning science is all about learning new causal principles: buoyancy, inertia, plate tectonics, electric current, homeostasis, respiration, photosynthesis, viral infection, or sexual selection. Some principles are specific to the phenomena they explain, whereas others embody forms of causation that cut across multiple phenomena. Feedback loops, for instance, regulate everything from metabolism to stock markets to climate change. Opportunity costs constrain activities as diverse as parenting, policy making, and product design. And selection pressures fuel not just evolution but also consumer habits and cultural transmission. New causal principles can unite spaces of possibility that were previously viewed as distant and discrete.

A prime example is the principle of emergence.[11] Some large-scale patterns, like predator–prey cycles, are caused by random interactions among smaller elements. The proportion of predators to prey remains stable in most ecosystems because when predators eat prey more quickly than the prey reproduce, the predators begin to starve. With fewer predators, the prey then rebound, taking over the environment. But an environment full of prey can sustain more predators, who rebound themselves. This tug-of-war between predators and prey keeps both populations in check without any plan or foresight on behalf of those involved. Predators do not intentionally curb their appetite, nor do prey intentionally curb their reproduction. They keep each other in check through their collective interactions.

Emergent processes have many properties that conflict with our intuitive understanding of objects and agents.[12] These processes occur across an entire system and cannot be localized to the elements within. They move toward equilibrium, such as a consistent ratio of predators to prey, but they have no beginning or end because the elements within the system continue to interact, simultaneously and independently. The system is in constant flux, yet patterns emerge nonetheless. Storms emerge from the collective interactions of air particles; flocks emerge from the collective interactions of migrating birds; markets emerge from the collective interactions of merchants; cities emerge from the collective interactions of builders; consciousness emerges from the collective interactions of neurons.

The principle of emergence can serve as a powerful new way to understand phenomena that defy agent-based or object-based explanations. Students who are taught the fundamental properties of emergent processes are generally able to apply those properties from one system to another.[13] If, for instance, students learn how the diffusion of a gas is systemwide, simultaneous, equilibrium-seeking, and ongoing, they can then apply these same properties to electric currents by noting how the interaction of electrons resembles the interaction of gas particles. Such instruction has proven beneficial to learners of all ages and can produce long-lasting improvements in scientific reasoning.[14]

Emergence is one of many unifying principles in a good science curriculum. Such principles, if taught appropriately, can reorganize the way we imagine natural phenomena.[15] The typical science curriculum is organized by content domain, such as chemistry, biology, and physics, but it might better be organized by the causal principles that cut across those domains. Consider the following three passages, which could be grouped by content or causality:

- A thermostat works by measuring temperature and turning on or off a furnace or air conditioner to reach a desired temperature. If the temperature is too cold, the thermostat will turn on the furnace until it becomes warm enough. Likewise, the thermostat on an air conditioner turns on when the house is too warm.
- Internet routers work by distributing a data signal to multiple devices. If the router is turned off, all the computers lose the signal. However, the functioning of one individual computer does not affect the functioning of the other computers. Thus, if one computer is turned off, all the others still get the data signal from the router.
- The Federal Reserve has the ability to raise or lower interest rates depending on the current state of the economy. If the economy is slow, the Fed will lower interest rates to stimulate borrowing and economic activity. Raising interest rates will slow the economy by increasing the cost of borrowing.

The first two passages are about electric devices whereas the third is about a government agency, so we tend to judge the first two passages as most similar. But the third passage and the first passage share a deeper commonality: they both describe systems regulated by negative feedback. Both thermostats and the Federal Reserve produce actions designed to counter prevailing conditions. Although the conditions they regulate are quite different (air temperature versus economic growth), as are the actions they bring about (heating or cooling a room versus raising or lowering interest rates), the underlying patterns of causality are the same.

These passages were taken from a set of twenty-five given to science experts and science novices in a study of categorization.[16] The participants were allowed to sort them in whatever way made sense, and both groups ended up creating five general categories. But the nature of those categories were vastly different. The novices sorted the passages by whether they pertained to biology, economics, environmental science, electrical engineering, or mechanical engineering, whereas the experts sorted them by whether they described a negative feedback loop, a positive feedback loop, a set of common causes, a set of common effects, or a linear causal chain. Science experts have a broader set of principles for thinking about causation. These principles connect seemingly disparate phenomena within the landscape of causal imagination and may even reorganize the landscape itself.

The Magic of Mechanism

Causal principles come in different forms, some more powerful than others. The principle "rain causes rainbows" is useful to know, but it doesn't provide any unique insight about rain, rainbows, or how they are related. We can use this principle to predict when a rainbow might occur, explain the presence of a rainbow, or create a rainbow of our own (using a sprinkler). But to answer deeper questions about rainbows—why they are rare, why they appear as an arc, why they contain the same colors—we need to know the mechanisms behind their generation. We need to know more about light, refraction, and vision.

Mechanisms differ from causal generalizations, like "rain causes rainbows," in three ways: they operate at a different level of analysis from the events they explain, they appeal to processes specific to the event, and they assume a necessary connection between the event's causes and effects.[17] Take, for instance, a mechanistic explanation of rainbows as optical illusions produced by the refraction of light through water droplets. This explanation appeals to processes involved in generating rainbows rather than those merely associated with them. These processes are specific to rainbows; they couldn't be applied to lightning, tornados, or sunsets. And they specify the conditions necessary for rain to produce a rainbow. Rainbows are rare not because the connection between rain and rainbows is inherently capricious but because many factors—the light, the moisture, our eyes—have to align just so.

When we observe an association between two events, such as rain and rainbows, we don't just store the association in memory but try to make sense of it through the mechanisms we know. Many people are unaware of the optical mechanisms behind rainbows, so they make sense of them in other ways, such as an accumulation of particles or a change in the quality of the air. Rarely are we content to learn the parameters of an association without also trying to explain it.

Indeed, studies have shown that when we seek to explain an event we ask questions about mechanisms, not associations.[18] If we happen to be given information about associations—how one factor covaries with another—we tend to interpret that information through the lens of potential mechanisms.[19] If, for instance, you learn that a friend was recently involved in a string of car accidents, you wouldn't conclude that she is simply accident prone. You would try to find out why. Is she having trouble concentrating? Is her eyesight impaired? Has she been drinking and driving? When you do find evidence of a possible mechanism—that your friend was, say, driving without her glasses—you'll likely cling to that mechanism and dismiss data implicating other possible factors, such as heavier than usual traffic.

Our focus on mechanisms often transcends our understanding of them, particularly when trying to explain physical systems or natural phenomena, but we still have reliable intuitions about their nature and complexity. For instance, few people understand the mechanisms behind a flashlight, a radio, or a microscope, but most people think that radios are more complex than flashlights and microscopes are more complex than radios. Likewise, few people know the mechanisms behind teeth, tongues, or hearts, but most people think tongues are more complex than teeth and hearts are more complex than tongues. Complex mechanisms are assumed to underlie systems with complex functions.[20]

This assumption constrains how we acquire causal knowledge, as well as how we evaluate the causal knowledge of others.[21] Suppose you were in the market for an electric car, and two of your friends just bought one (different friends than the friend who gets into accidents). Your friend Troy based his decision on the car's shape, color, and interior. Your friend Abed bought the same car but based his decision on the car's speed, horsepower, and battery. Who seems to know more about electric cars? Who would you consult for advice on your own purchase?

Clearly, you would talk to Abed. Abed not only seems to understand the mechanisms of electric cars but may also know more about cars in general. Even kindergarteners agree. They would ask Abed for advice over Troy, despite knowing little about cars themselves.[22]

Mechanistic information provides a depth of understanding that mere factual information does not. Indeed, mechanisms allow us to interpret facts that would otherwise defy interpretation. A prime example comes from the domain of evolution. People notoriously misunderstand evolution, viewing it as the uniform transformation of a population rather than the selective survival and reproduction of a subset. That is, most people think of evolution as a kind of cross-generational metamorphosis, whereby monkeys metamorphosed into apes and apes metamorphosed into humans. But evolution actually involves the culling of nonadaptive traits from a population and the spread of more adaptive ones.[23]

One reason we misunderstand evolution is that for many years we are taught about biological phenomena without being taught the mechanisms behind those phenomena.[24] Variation, inheritance, predation, parasitism, altruism, mimicry, and the like are taught as isolated facts, disconnected from each other and from the processes that shaped them. The logic that unifies these facts—evolution by natural selection—is left out of the elementary school curriculum because it is viewed as either too controversial or too complex. But elementary schoolers can indeed learn the logic of natural selection. They can learn it as readily as adults, and they can transfer that logic from one instance of adaptation to another.[25] They can even learn how natural selection in combination with geographic isolation gives rise to new species.[26]

Natural selection is a causal mechanism that expands the biological imagination far and wide. It provides a means of unifying known biological facts while also forging paths to novel biological ideas like those discussed at the beginning of the chapter. And, for these reasons, people who understand natural selection are also more likely to accept evolution as true.[27] Evolution remains one of science's most controversial ideas, mostly because it contradicts religious teachings, but people who understand how evolution works tend to accept that evolution occurs.

The same holds for another controversial scientific idea: climate change. People who understand the main mechanism behind climate change—that greenhouse gases trap heat from the sun—are more likely to accept that the climate is actually changing and that humans are the cause.[28] What's more, mechanistic knowledge predicts acceptance across the political spectrum for both evolution and climate change.[29]

Mechanistic knowledge thus transforms our relationship to ideas that stretch imagination, deepening our understanding of these ideas and increasing our acceptance. It can also change our behavior.[30] For example, many people think that colds are caused by being cold, typically through exposure to wind, rain, or snow. Thus, to prevent colds, many people focus on avoiding cold weather rather than avoiding the real cause of colds: germs. The idea that invisible organisms invade our bodies and consume our cellular resources defies imagination. But studies have found that disease-prevention programs that focus on germs and the mechanisms behind their survival and reproduction lead to better disease-prevention behavior, such as sanitizing one's hands before handling food.[31] Mechanistic knowledge helps us parse genuine associations, such as the association between germs and disease, from spurious ones, such as the association between disease and cold weather, which, in turn, helps us prioritize productive behaviors over counterproductive ones.

The Art of Abstraction

We have discussed how causal principles expand imagination by connecting and organizing disparate facts and how mechanisms expand imagination by adding depth and meaning to those connections. Abstraction is the final ingredient, allowing us to transfer causal mechanisms from one situation to another. Abstract causal mechanisms are the holy grail of science as well as science education, but their discovery is hampered by our tendency to focus on specific objects and specific situations. Applying general ideas to specific situations can be hard because the devil lurks in the details. But working in the opposite direction, from specifics to generalizations, is much harder; the devil simply is the details.

The process of abstraction is really several processes: noting commonalities between two situations; aligning the elements of those situations so that the objects in one situation correspond to objects in another; shifting attention from common objects to common relations; noticing places where the relations in one situation do not map onto the relations in the other and then adjusting the mapping, and finally, describing the common relations across situations in general terms, thereby abstracting a schema that pertains to both.[32] Comparison facilitates analogy, and analogy facilitates abstraction.

This description of abstraction is itself abstract. As an illustration, let's consider physicist Ernest Rutherford's discovery of atomic orbitals, as described by psychologist Dedre Gentner.[33] At the beginning of the twentieth century, Rutherford investigated the structure of the atom and noticed commonalities between atoms and the solar system. The nucleus of an atom seemed to correspond to the sun, and its electrons seemed to correspond to planets. This correspondence alerted him to several common relations: electrons are smaller than the nucleus just as planets are smaller than the sun; electrons are attracted to the nucleus just as planets are attracted to the sun; and electrons orbit the nucleus just as planets orbit the sun.

Rutherford didn't know why these relations are true of atoms, but he did know why they are true of the solar system: the difference in mass between planets and the sun causes the planets to orbit the sun, as their paths are perpetually curved by the sun's gravitational pull. He inferred that the same must be true of atoms—that electrons orbit the nucleus because they are perpetually pulled toward the larger body.

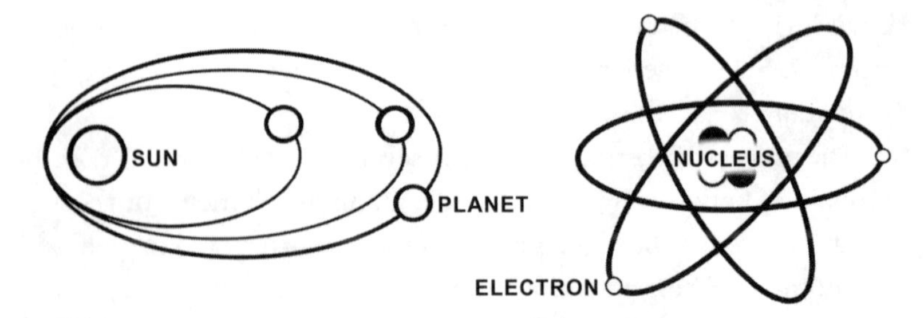

Figure 5.1 Analogies are the engine of scientific discovery, such as the discovery that electrons revolve around the nucleus of an atom similarly to how planets revolve around the sun. (Courtesy of the author)

We now know the forces at work within an atom are different from those at work within the solar system, but Rutherford's insights about orbital mechanics paved the way for modern chemistry. Rutherford discovered something genuinely novel by drawing an analogy between two sets of facts he already knew. Discovery by analogy is common in the history of science. Johannes Kepler discovered the principles of celestial motion by comparing gravity to light.[34] William Harvey discovered the principles of circulation by comparing blood flow to the water cycle.[35] Darwin discovered the principles of evolution by comparing population growth to the growth of human economies.[36] And the Wright brothers discovered the principles of human-powered flight by comparing airplane wings to bird wings.[37] We typically think of science as driven by empirical discovery, but many scientific breakthroughs were driven by a reanalysis of things already discovered—a comparison of known facts within the workspace of imagination.

The reason analogy spurs science is that it reveals causal mechanisms that go unnoticed when attending to a single system but become apparent in the comparison of two seemingly distinct systems. Such discoveries can be as potent for science students as for practicing scientists. Analogizing the force a table exerts on a book to the force a spring exerts on a hand can help students appreciate the principle of balanced forces.[38] Likewise, analogies between boiling water and churning magma can help students appreciate the principle of thermal convection, and analogies between human vertebrae and giraffe vertebrae can help students appreciate the principle of common descent.[39]

Part of the reason analogies are effective in science instruction is that we like analogies. We make them easily and often, especially when communicating with others.[40] Analogies are an effective way to highlight a key feature of a situation, such as a cause–effect relation, and convince others that the feature matters. Observational studies of scientists and doctors have found they make as many as fifteen analogies per hour when problem-solving with their colleagues.[41] Novice problem solvers also make analogies, though only around four per hour.[42] Even children analogize during collaborative problem-solving. One study found that fourth graders generated around five analogies per collaboration, including such gems as "this classroom is like a cage" and "it's like you're painting outside in the rain."[43]

But just because we like analogies doesn't mean we're skilled at making them, particularly when analogizing situations with seemingly little in common. Things can go wrong at each step in the analogizing process, and they frequently do. The first step is to align common elements across situations, and we can become distracted by their superficial features.[44] Recall that in Rutherford's analogy between an atom and the solar system the informative alignments were nucleus-to-sun and electrons-to-planets, but what if Rutherford had focused on the fact that the sun has a lot of energy and so do electrons? He might have aligned electrons to the sun and discovered no meaningful relations therein.

Beyond aligning objects, we can falter in finding ways to compare them. Superficial features continue to pose a threat because they distract us from deeper, more meaningful commonalities. The more we attend to the features of the objects we are comparing, the less we notice the relations among them.[45]

An apt example of this problem comes from a developmental study where children were shown a collection of carrot gardens, some with healthy carrots and some with sick carrots, and they were asked to determine which factors help the carrots grow.[46] Each garden had a unique feature—rocks, a doghouse, a tree—but some gardens shared features, such as the type of soil. In fact, all the gardens with healthy carrots had red soil (as opposed to brown), but children typically failed to notice this commonality. When asked to describe where carrots grow best, they listed the gardens' unique features and ignored their common one. Adults are prone to the same error, especially when encountering analogous situations without any prompt to compare or explain them.[47]

Comparing distinct situations is necessary for discovering a common principle but not sufficient; the final step is abstracting a general description of the principle that transcends the particulars of the original situations. This step requires active analysis and does not come for free simply by considering multiple instances of the same principle.

Recall the task in which participants learn about thermostats, internet routers, and the Federal Reserve and are asked to identify the two systems that are most similar. The response based on abstract principles would be to group thermostats with the Federal Reserve because both operate on the basis of feedback, but nonscientists tend to group thermostats with internet routers because both are electric devices. Convincing nonscientists to make the first grouping is quite difficult. Pointing out that both the thermostat and the Federal Reserve involve feedback is not particularly helpful, nor is providing a description of how each of those systems embodies the principles of a feedback loop.

The only instruction that reliably changes how participants conceptualize these systems is asking them to write a description of what thermostats have in common with the Federal Reserve. This activity forces them to note commonalities at a higher level of abstraction and, consequently, to ignore the superficial details that led them to group thermostats with internet routers.[48] Similar results have been observed in other domains of knowledge, including algebra, logic, economics, and evolution.[49] People who are taught new principles in the context of specific examples learn those principles best when they can describe them at a level that transcends the examples themselves. Imagination benefits from abstraction but deals, most naturally, in specifics.

To recap, learning abstract principles through analogy requires aligning the right objects, making the right comparisons, and encoding those comparisons at the right level of description. But that's not all. Abstract principles are useful only if we can apply them to new situations, beyond those that gave rise to their discovery, and this task is complicated by its own challenges. The more remote the new situation—in time, construal, or context—the less likely we will apply the principle.[50] Rather than use the principle to carve new paths into the landscape of ideas, we may shelve it at the back of our minds.

Abstract principles, like other tools, require practice. The more we practice using them, the more easily and effortlessly we deploy them, as discussed in the next section. But there are other factors that can help us transfer principles from known situations to novel ones, such as learning tangible procedures for applying them. For instance, students who learn abstract mathematical principles as algebraic formulas are more successful at applying those principles to novel problems than students who learn the same principles in words, couched in the language of the training examples.[51] The abstractions inherent in algebraic equations may seem confusing or off-putting on their surface, but they are a blessing in disguise; they facilitate transfer better than principles grounded entirely in examples.[52]

Another factor that facilitates transfer is labels. Labels reify the principle and provide a tangible means of accessing it. The label "sexual selection," for instance, is much handier than the sequence of events it summarizes—namely, how mating preferences change the frequency of desired traits from one generation to the next by increasing the reproductive success of organisms who possess the trait at the expense of those who do not. The label stands in for all that machinery and provides an efficient means of discussing it with others. Labels like "symbiosis," "conservation of energy," and "supply and demand" are shortcuts for contemplating and discussing a complex arrangement of causal processes.

Labels also facilitate the learning of abstract principles, highlighting common relations among distinct examples while providing a framework for organizing and remembering those examples.[53] Carpool arrangements and free trade agreements seem to have little in common until thought of as two forms of reciprocity. Youth antidrug programs and medical vaccines seem to have little in common until thought of as two forms of inoculation. Labels turn abstract insights into tangible categories. A rose by any other name may smell as sweet, but an abstract principle needs to be described by the same name to have the same inferential power.

Principled Perception

Abstraction allows a principle to be applied to multiple situations, but a principle cannot be wholly abstract. We need to have ways of identifying situations to which the principle applies—ways of perceiving situations in terms of features relevant to the principle. This special kind of perception comes with practice. The more we apply a principle, the more easily we perceive the fit between it and the world. The cognitive changes brought about by using the principle are accompanied by perceptual changes in applying the principle.

These perceptual changes are a hallmark of expertise. Experts analyze the world for the affordances that will allow them to apply the principles they know best.[54] They attend to patterns that correlate with those observed many times before, and they encode them, in their minds, in terms of the principles that organize their expertise. They then use these encodings to further analyze the situation, searching for information predicted by the relevant principles but not yet observed. When they discuss their observations with others, they discuss them as abstractions, devoid of the concrete details that led to those abstractions, and they remember them as abstractions as well.

Consider the perceptual skills of expert geologists. Geologists don't just see rocks; they see morphological transformations. They analyze patterns on the surface of a rock for evidence of large-scale, systematic changes in the earth's crust. These analyses concern processes that cannot be perceived—changing climates, shifting plates, pooling magma—but features of these processes can. And to interpret the features appropriately, geologists employ a variety of skills, including mental rotation, geometric projection, and spatial decomposition.[55] Geologists can tell, by sight, the difference between a fracture, a fault, and a fold. They can distinguish the effects of different geological events, such as landslides, earthquakes, and erosion. And they can visualize the contours of a rock structure embedded inside other structures.

Psychologists have investigated the spatial reasoning of geoscientists, including their navigation skills, visualization skills, and location memory, and they find that geoscientists outperform other scientists on measures of these abilities.[56] They even outperform chemists, who also analyze spatial patterns as part of their job but not the same patterns. One of the most informative differences between geoscientists and chemists comes from a faulted-word task, where participants read words that have been deformed along a diagonal "fault" line. Although neither group has experience with

CONCRETE

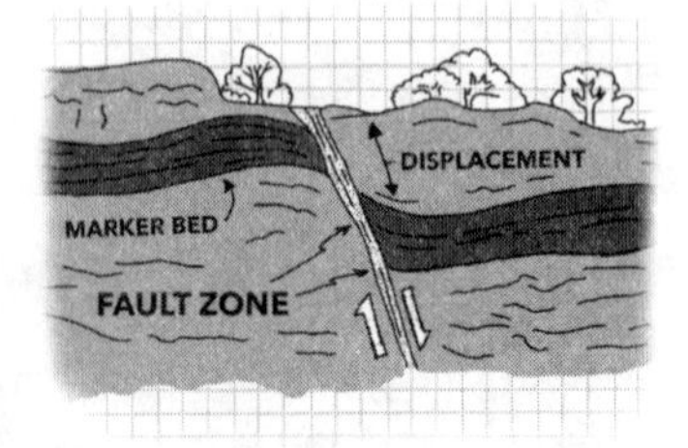

Figure 5.2 Scientific principles change the way we perceive the world, as when geologists perceive rock formations in terms of the forces that produced them. (Adapted from Jee and Anggoro 2021) (Courtesy of the author)

this task, geoscientists read the words more accurately than chemists, presumably because they recruit their expertise at reading faulted rocks to realign the fragments of the faulted words.[57]

Experts in other domains develop unique perceptual abilities as well. Painters analyze snapshots of natural scenes for their three-dimensional layout, whereas architects analyze the same scenes for their paths and spatial boundaries.[58] Mathematicians parse visual diagrams into geometrically relevant configurations, such as congruent angles, parallel lines, and right triangles.[59] Expert Tetris players perceive the Tetris screen not as a collection of blocks but as higher-order structures like pits and wells. And these structures trigger not just single actions but action sequences, composed of precisely timed translations and rotations.[60]

The more we analyze a domain in terms of abstract principles, the better we perceive the signatures of those principles. Analysis via principles changes what biologists call the *umwelt,* or the unique way an organism perceives its environment. Every organism is an expert at survival, exquisitely adapted for life within a particular niche, and this expertise is facilitated by unique forms of perception.[61] Bats are experts at hunting insects at night because they perceive flying insects through reflected sound.

Scorpions are experts at hunting rodents in the desert because they perceive scurrying rodents through vibrations in the sand. Sharks perceive bioelectric fields to track swimming fish; bees perceive ultraviolet light to find flowers marked with ultraviolet patterns; and geese perceive magnetic fields to maintain their bearings during flight.

In this same vein, humans use their perception of abstract properties to achieve goals that would otherwise be unachievable. Geologists use their perception of faults and folds to determine the evolution of a rock structure; mathematicians use their perception of parallel lines and congruent angles to prove theorems; and Tetris players use their perception of pits and wells to earn more game points. Yet a marked difference between the human umwelt and the umwelt of other animals is that we can change our umwelt. We can restructure our own perception. We may not be able to develop new sensory receptors, like the electroreceptors of sharks or the magnetoreceptors of geese, but we can fine-tune the receptors we do have to identify features aligned with new goals and, hence, new possibilities.

While other animals are stuck seeing the possibilities they evolved to see, we can see possibilities far beyond those boundaries if we learn the right principles. Principles change how we perceive the world because they change how we conceive it—how we interpret it, describe it, and organize it. In science, the relevant principles are typically causal mechanisms, but mathematics principles and ethical principles can be equally transformative, as discussed in the following chapters. Principles may connote monotony and rigidity, but they actually inspire novelty and creativity. A principled imagination is a productive imagination.

6 Mathematics

Expanding Our Numerical and Spatial Imagination

Let's ponder for a moment the curious connection between odd numbers and square numbers. If you add one and three, you get four, which is a perfect square, the square of two. If you add one, three, and five, you get nine, another perfect square, the square of three. Add seven to this sum and you get sixteen, the square of four. Add nine and you get twenty-five, the square of five. The sum of consecutive odd numbers is a perfect square for numbers one through nine, but does this pattern hold for larger numbers? Would the sum of odd numbers from one to ninety-nine be a perfect square? What about the sum from one to one trillion and one?

Yes, the sum of consecutive odd numbers is always a square number. To see why, imagine each square number as a geometric square. Four could be imagined as four blocks, stacked two by two, and nine could be imagined as nine blocks, stacked three by three. If you started with a square of four blocks and wanted to make a

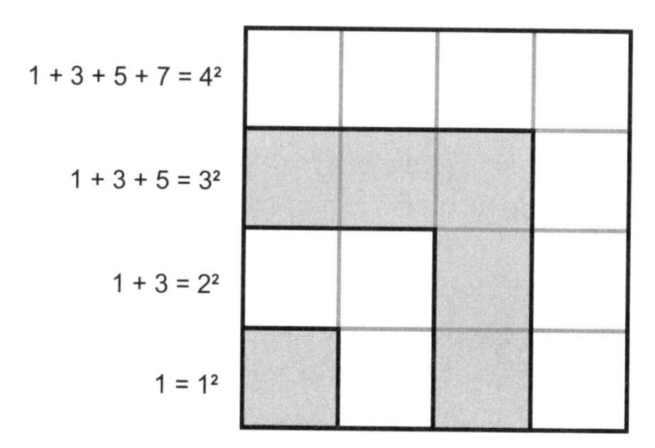

Figure 6.1 The sum of consecutive odd numbers is always a perfect square. The reason can be visualized in terms of actual squares. The number of one-unit squares needed to increase a square's perimeter is always one more than twice that square's length. (Adapted from Johnson and Steinerberger 2019) (Courtesy of the author)

square of nine, you would add one block to each row of two, one block to each column of two, and one block to fill the gap between them—a cornerstone completing the three-by-three arrangement and bringing the total number of blocks to five. If you started with a square of nine and wanted to make a square of sixteen, you would add one block to each row of three, one block to each column of three, and one block to fill the gap, for a total of seven.

The number of blocks needed to move from one square to the next is one more than twice the length of the original square, which guarantees that it will be odd. And this number will not only be odd but also the next consecutive odd number, because increasing the size of the square requires exactly two more blocks than required for the previous increase.

This peculiar relationship between odd numbers and square numbers was discovered by the seventeenth-century mathematician Johann Faulhaber. The math behind Faulhaber's formula is formidable, but the geometric proof—of modeling square numbers as actual squares—is intuitive. In fact, people find this proof not only intuitive but beautiful.[1] They rate it as elegant, profound, and clear, similar to how they rate their experience of a painting or a symphony. Mathematical ideas have the power to stir our aesthetic emotions, even if our knowledge of mathematics is limited to what we learned in high school.

The geometric argument for why consecutive odd numbers sum to square numbers may seem intuitive, but it involves ideas that are not part of early intuition. A child who has just learned to count knows nothing of odd numbers, square numbers, addition, or division. They do know that odd numbers follow every even number; that odd numbers cannot be evenly divided by two; or that the smallest odd number is one but the largest odd number is undefined. They do not know how to multiply two numbers together or how to divide one number by another. They do not even know that numbers correspond to distinct quantities. The vast machinery of mathematics—from integers to integrals, decimals to decahedrons—must be built from scratch.

It may seem obvious that we have to learn math, but what is less obvious is that such learning occurs largely within the realm of imagination. We do not find mathematical entities in the world, waiting to be perceived and explored. We invent them, first as a cultural innovation and then as an educational achievement. Mathematical entities describe what we perceive, but they are not a consequence of perception. People can (and do) live their entire lives without inventing the concept of number, let alone specific types like odd numbers and square numbers.

In this chapter, we'll discuss how it's possible to live one's life without a precise sense of number, but why most people today do, in fact, develop that sense. More generally, we'll discuss how we discover and develop key mathematical ideas, from new types of numbers to new mathematical operations. Learning about numbers goes hand in hand with learning about operations. New operations reveal the existence of new numbers, and new numbers are needed to fully apply new operations. Operations are the principles that expand mathematical imagination; how we learn them provides a particularly pure example of how principles open new landscapes of possibility.

From Manipulations to Operations

There are several types of number, and each is defined by a distinct operation.[2] If you don't know the operation, you can't imagine the number. Counting gives us whole numbers but not integers because subtraction is required to define negative numbers. Subtraction gives us integers but not rational numbers because division is required to define fractions. And division gives us rational numbers but not real numbers because roots are required to define irrational numbers. Any whole number (3) is also an integer (+3), a rational number (3 / 1), and a real number (3^1), but these additional senses must be constructed.

Constructing a new sense of number is a process of abstraction. Whole numbers are abstracted from grouping objects into sets. Integers are abstracted from ordering whole numbers along a number line. Rational numbers are abstracted from entering integers into proportions. And irrational numbers are abstracted from dissecting rational numbers into roots. The increased abstraction of numerical concepts allows reasoners to apply them broadly and efficiently to any object (dollar, dog, drink, or dream) and any means of manipulating those objects (cutting, cropping, combining, or compounding). But such benefits come at the cost of transparency. The connection between mathematical concepts and the situations they describe becomes increasingly opaque and may vanish altogether.

Consider the following problem: "Lucille won some money in a lottery. She kept $64 for herself and gave each of her three sons an equal portion of the rest. If each son got $21, how much did Lucille win?" If you've taken basic arithmetic, then you can probably solve this problem without difficulty. You'd multiply the $21 that each son received by three, and then add that product to the $64 that Lucille kept for herself, yielding a total of $127. You could also solve this problem using algebra. You could let X stand for the amount Lucille won in the lottery and write an equation that relates X to the other quantities, namely, $(X-64) \div 3 = 21$.

Representing the problem this way adds a level of abstraction that ends up being more harmful than helpful.[3] We are more likely to make calculation errors, such as dividing sixty-four by three rather than multiplying both sides of the equation by three. The terms in the equation are no longer connected to the concrete object of dollars and the concrete procedure of distributing, and we start making errors we would never make when thinking about the real-world situation described by the equation. Why, for instance, would we ever divide Lucille's share of the money by three?

Representing concrete situations with abstract equations is not always a problem, however. Consider this problem: "Levi just paid $34 for new jeans. He got them at a 15 percent discount. What was the original price?" Most of us would solve this problem incorrectly. Our inclination is to multiply the price Levi paid by 0.15 and then add that product to the price, but that calculation does not give us the original price of the jeans. It gives us the meaningless sum of the discounted price plus a second iteration of the discount.

To get the jean's original price, we need to divide the discounted price by 0.85, which is not at all intuitive. It only becomes intuitive with some algebra. If we let X stand for the jean's original price, we can represent the discounted price as $X - 0.15X = 34$. We can then simplify the equation to $0.85X = 34$ and solve for X by dividing both sides of the equation by 0.85. This procedure reveals that the original price of the jeans was $40—an answer we are more likely to obtain when given the equation than when given a verbal description of the problem.

These two problems illustrate the trade-off between physical manipulations and abstract operations. Distributing a resource, like money, is a well-practiced task, easily understood within the context of who is distributing the resource and how. Such details help us perform the appropriate calculations and avoid senseless errors. Inversing a discount is not a well-practiced task, and the details surrounding this task provide little support for carrying it out. Inversing a discount requires relating the unknown quantity to itself, and this relation is easier to understand when the quantity is abstracted away from the context, as variable *X*. The more complex the operation, the more difficult it is to model in imagination and thus the more we stand to benefit from abstraction.

Research on the instructional benefits of manipulatives, like blocks or tokens, has revealed similar findings. Manipulatives are routinely used in the classroom as a way of turning abstract concepts into tangible experiences, such as when students learn place value by manipulating base-ten blocks or when they learn fractions by manipulating pieces of a pie. But studies have found that manipulatives have limited practical value beyond introducing a new concept. They help students remember the concept, but they are less helpful at facilitating problem-solving and virtually no help at transferring solutions from one problem to another.[4]

Students who learn fractions as pieces of a pie, for instance, will remember that the numerator corresponds to the number of pieces you have and the denominator corresponds to the size of those pieces, but they'll struggle to order fractions and add fractions, and they'll struggle even more when working with fractions that do not correspond to the predivided pies they originally manipulated.[5]

Concrete objects constrain our thought and, accordingly, our imagination. Even after we've mastered abstract mathematical operations, we continue to think of them in terms of the physical manipulations they best exemplify. Division, for instance, is most easily understood as cutting or distributing, as when we divide a cookie in half or divide a collection of cookies into groups. We divide cookies by the number of people who want to eat them. But we could also divide people by cookies, distributing groups of people to each cookie or even parts of people. We think of the process of dividing six cookies among three friends as distributing two cookies to each person, but we could also think of the process as assigning half a person to each cookie.

The abstract nature of division, as a mathematical operation, makes both interpretations possible, but we prefer the interpretation that accords with the physical manipulation it models. Textbooks illustrate division with situations that involve distribution, like dividing tulips among vases, whereas they illustrate addition with situations that involve combination, like adding tulips and daisies. Rarely do they ask students to add tulips and vases or divide tulips by daisies.[6]

In fact, when presented with physically incongruous problems like "divide 12 tulips by 4 daisies," we make more errors than when presented with congruous versions the same problem, like "divide 12 tulips by 4 vases." And while solving physically incongruous problems, our brains elicit neural signals associated with error detection and surprise.[7] A correct yet incongruous statement, like "12 tulips divided by 4 daises is 3," elicits the same signature of surprise as a congruous yet incorrect statement, like "12 tulips divided by 4 vases is 5." Incongruity between an abstract operation and a physical manipulation can trip up everyone, even skilled mathematicians.[8]

Mathematical imagination requires abstraction to solve novel problems with familiar operations and use familiar problems to devise novel operations. But abstraction is difficult to achieve when learning and difficult to maintain when reasoning. In the following sections, we'll examine three cases of deriving abstract concepts from concrete procedures: deriving integers from counting and adding, deriving fractions from comparing and dividing, and deriving geometry from rotating and inverting. Each case illustrates how abstract operations expand our understanding of quantity and space while also illustrating how foreign those operations are to our earliest conceptions of mathematical possibility.

A Sense of Precision

In contemplating the origin of mathematics, Leopold Kronecker, a nineteenth-century mathematician, said that "the integers were created by God; all else is the work of man."[9] Complex mathematical ideas like decimals, quadratics, and logarithms clearly seem to be inventions of mathematical ingenuity, but, surprisingly enough, so are integers. Knowledge of integers is not innate but rather constructed from the operations of counting.

Before we learn to count, we perceive number as a feature of the environment but only approximately. We can tell, for instance, which of two apple trees has more apples, but we can't tell exactly how many more. Our comparison relies on a general sense of magnitude rather than a precise value—the total or "cardinal" value of the set. A collection of thirty apples would appear more numerous than a collection of twenty, but that perceptual difference is akin to how a shiny apple appears brighter than a dull apple or how the thud of several apples hitting the ground sounds louder than the thud of a single apple.[10] The cardinal difference between two sets of apples cannot be measured without a system for tagging and grouping individual apples. We might be able to distinguish twenty apples from thirty, but we couldn't distinguish twenty-nine from thirty or even 120 from 130. Our perceptual sense of number is too fuzzy, and that sense gets fuzzier as the numbers get larger.[11]

As an illustration, try tapping your finger on your leg twenty times without counting. You might make twenty taps exactly, but you might also undershoot a bit, making eighteen or nineteen taps, or overshoot a bit, making twenty-one or twenty-two taps. If you repeated this activity several times—and with the help of a partner who could count your taps—you would produce a distribution of taps centered at twenty but with some error on either side. If you did the same task but aimed to tap your finger thirty times, you'd produce a distribution of taps centered at thirty but with more errors. You'd miss thirty by a few taps more often than you'd miss twenty by a few taps. That's the fuzzy logic of estimation.

Estimation skills turn out to be ubiquitous across the animal kingdom. Animals as diverse as rats, crows, squids, and bees are able to estimate a set's "numerosity" and compare sets of different numerosity, as revealed by studies in which animals are rewarded for producing a specific number of responses or for selecting the larger of two sets.[12] Estimation is ubiquitous across development as well. By six months, human infants can tell which of two sets is larger, if the sets vary by a ratio of at least two to one, and they can do so for many kinds of sets, including collections of toys, arrays of dots, sequences of tones, and series of actions. As we age, we refine these estimation skills, using them to compare sets of increasingly smaller ratios and for sets that are too large or too fleeting to count.[13]

Perceptual estimation is ubiquitous because the need to make comparisons is ubiquitous; all organisms need to discern where resources are most abundant and threats are least abundant. But comparison is the only operation fully supported by estimation. We cannot use our fuzzy sense of number to keep track of how many years have passed since we were born, how many people live in our community, or how many meals our neighbor owes us. We cannot even use it to entertain whole numbers, like twenty. Twenty is not just an amount somewhere between eighteen and twenty-two; it's a set of items exactly one more than nineteen and exactly one less than twenty-one.

As skilled counters, we take the power of whole numbers for granted. We forget how impoverished our numerical imagination was without them. Children who are just learning to count—and thus just learning about whole numbers—make surprising errors when applying their count words to sets. If you tell them that you have eight apples and your friend has four, they are unsure who has more.[14] If you show them a picture of eight apples and ask them to find another picture of eight, they are more likely to match apples by color than by number.[15] And if you show them a tray with eight apples and then remove one of the apples, they are unsure whether the tray now contains seven apples or nine.[16] Some even claim it still contains eight.[17]

These errors underscore the fact that counting is more complicated than it looks. Counting has a logic that children learn to appreciate slowly, in steps.[18] First, children learn their language's count list, or list of words used to denote sets of increasing numerosity ("one," "two," "three," and so on). Second, they learn how to apply this list to a set of objects by labeling each object with one, and only one, word in the count list. Third, they learn that, when applying the count list, the last word they reach corresponds to the cardinal value of the set. "Eight," for instance, refers not to the eighth object tagged but to all objects tagged, collectively.

We know that children learn these skills in stages rather than in tandem, because of the gaps between them. Children learn to recite the count list long before they can apply it to a set of objects without double-counting, undercounting, or skipping numerals.[19] And they learn to apply the count list long before they realize that the last word reached answers the question, "How many are there?" In fact, children go through a prolonged period of being able to count a collection of objects without being able to retrieve a particular number of objects from the collection. When asked to retrieve five marbles from a collection of ten, they grab a handful at random and make no attempt to coordinate their knowledge of counting with their assessment of numerosity.[20] Counting is initially understood as a ritualized activity, like reciting the alphabet or saying pattycake, and it continues to be understood as such for many months before children grasp its mathematical implications.

Of crucial importance to counting is exposure to a count list. Some cultures, like the Piraha and the Munduruku of the Amazon rainforest, do not have count lists, so the members of those cultures cannot keep track of exact numerosities.[21] The Piraha and Munduruku have words for "one," "two," "some," and "many," but they do not have words for large, precise quantities, like "eight" or "eighty-eight." If the Piraha or Munduruku watch someone put eight nuts in a can and are asked to do the same themselves, they will put approximately eight nuts in the can. Their mean response across many trials will be eight, but their response on any given trial might by six, seven, nine, or ten—similar to what would happen if you tried tapping your finger eight times without counting.

In the absence of a count list, the Piraha and the Munduruku rely on an approximate sense of number to compare quantities. The same has been observed for deaf adults who lacked exposure to language when they were children and, accordingly, to the words for counting.[22]

Counting is thus the first step in expanding numerical imagination. It does so not only by defining a precise sense of quantity but also by laying the foundation for defining new types of numbers. About two years after children learn to count, they hit upon two further insights: that every number can be increased by one and, accordingly, that there is no biggest number.[23] Counting thus paves the way for infinity. The more proficiently a child can count—hitting all the numbers and making the appropriate decade changes (from, say, forty-nine to fifty)—the earlier they recognize that counting goes on forever.[24]

Counting also paves the way for addition and subtraction. If we count by units greater than one, then we are performing addition, and if we count backward, then we are performing subtraction. Addition and subtraction are alternative ways of moving through the count list. They increase the efficiency and scope of our numerical comparisons, but they also challenge the boundaries of the count list itself. If we can subtract four from seven, then we can also subtract seven from four, yielding a negative number, −3. We might even subtract seven from itself, yielding zero.

Historically, it took humans centuries to discover numbers beyond the count list. Today, we are introduced to those numbers as children, but we still require many years to master them.[25] Negative numbers are particularly challenging. They are viewed not as reverse steps on a number line but as distinct kinds of numbers. Ordering negative numbers takes longer than ordering positive numbers of the same absolute magnitude, even for mathematically literate adults, and performing simple calculations with negative numbers, such as $7+(-4)$, takes longer than performing the same calculations recast as subtraction, such as $7-4$.[26] When negative numbers are involved in a calculation, we are more likely to make errors and less likely to spot errors that someone else has made.[27] Subtraction makes negative numbers possible, but we have to practice working with them to imagine them as magnitudes rather than mere symbols.

In many ways, negative numbers are the antithesis of the fuzzy numbers we know innately. Fuzzy numbers are approximate representations of real-world quantities that do not support precise mathematical operations, whereas negative numbers are products of precise operations that have no real-world counterpart, at least not one we can see or hear. Negative numbers are tools for expanding our perceptual sense of numerosity, as are "zero" and "infinity," as well as large positive integers like "eighty-eight." But these tools must be reintegrated with that perceptual sense to be thought of as genuinely numerical—as referring to numbers and not just calculations.[28] This same predicament occurs when we construct ratio-based, or "rational," numbers, such as fractions and decimals.

A Sense of Proportion

Pop quiz: Which is greater, 1 / 4 or 1 / 3? Although 1 / 3 is greater, you may have felt a pull toward 1 / 4 because four is greater than three. How about 0.05 and 0.1? Although 0.1 is greater, you may have felt a pull toward 0.05 because five is greater than one. Rational numbers like 1 / 4 and 0.05 are expressed with whole numbers yet signify quantities that are not whole. Fraction bars and decimal points turn whole numbers into ratios—numbers defined by division, rather than counting or addition. The introduction of a new operation makes these numbers genuinely novel and, as a result, genuinely difficult to imagine.

One part of this difficulty is the new notation. When children are first introduced to fractions and decimals, they either ignore the new notation or interpret it incorrectly.[29] For instance, when asked whether 0.8 is greater than 0.08000, they often choose 0.08000, reasoning that "the zeros in the front don't mean anything but it has zeros in the back and that means that it is eight thousand." And those who choose 0.8 sometimes do so for the wrong reason, such as that "0.8 is 1 / 8 of a whole."[30]

But notation is not the only difficulty. Even students who understand that fractions and decimals refer to ratios have trouble assigning appropriate magnitudes to these ratios, as revealed by where they place fractions and decimals on a number line. While children in the early stages of fraction learning recognize that 4 / 5 is closer to one than zero, they typically indicate that 16 / 20 (an equivalent ratio) is even closer to one. Equivalent ratios are perceived as nonequivalent when different whole numbers are used to express them.[31]

This "whole number bias" in assessing fraction or decimal magnitudes is never completely overcome. Adults with many years of mathematics education show the bias when deciding which of two fractions or decimals is greatest, as you were asked to decide earlier.[32] A statement like "1 / 5 is less than 1 / 3" take longer to verify than a statement like "3 / 7 is less than 5 / 7" because the whole number components three and five pull us in the right direction when they appear as numerators but the wrong direction when they appear as denominators. This bias holds even for people with PhDs in mathematics, who manipulate fractions and decimals on a daily basis.[33] It also holds in situations where we are fully aware of the fractional magnitudes involved, such as deciding which of two gambles to take. We may recognize that the odds of winning a 50:50 gamble are the same regardless of whether our chance of winning is 1-in-2 or 9-in-18, but we still prefer the latter gamble, reasoning that it's better to have nine chances than one.[34]

The root of our difficulty in assigning magnitudes to fractions and decimals is that we fail to appreciate the operation that produces them: division. Division is several steps removed from the more familiar operations of addition and subtraction. If we add the same number several times, as in $2+2+2$, we can represent that process as multiplication: 2×3. But if we subtract the same number several times, we can't represent that process as division.

Some instances of division can be represented as repeated subtraction; $6\div3$, for instance, can be represented as subtracting two from six three times. But the analogy falls apart when dividing by numbers that do not fit neatly into the dividend, such as dividing six by five. The fraction 6 / 5 is a division problem whose solution cannot be expressed as a whole number. Children initially learn to solve such problems with remainders—in this case, "1 remainder 1." But this strategy is a conceptual dead end; it provides no clue as to how 6 / 5 could be a number unto itself, with a unique magnitude and a unique place on the number line.

Children's initial conflation of division and subtraction leads them to the peculiar conclusion that numbers will disappear if divided repeatedly.[35] Consider this conversation between a third grader and a researcher probing the child's understanding of division as a mathematical operation.

> *Researcher:* Are there any numbers between 0 and 1?
>
> *Child:* No.
>
> *Researcher:* How about one-half?
>
> *Child:* Yes, I think so.
>
> *Researcher:* About how many numbers are there between 0 and 1?
>
> *Child:* A little, just 0 and half, because it is halfway to one.
>
> *Researcher:* Suppose you divided 2 in half and got 1 and then divided that number in half. Could you keep dividing forever?
>
> *Child:* No, because if you just took that half a number, that would be zero, and you can't divide zero.
>
> *Researcher:* Would you ever get to zero?
>
> *Child:* Yes.

This child denies that numbers can be divided indefinitely, as well as that the number line contains numbers other than whole numbers. When pressed, he acknowledges that one-half falls between 0 and 1 but then denies the existence of any other fractional numbers.

As children learn more about fractions, particularly unit fractions like 1 / 3 and 1 / 4, they accept that the number line is more densely packed than they first imagined, but they continue to deny that numbers can be divided indefinitely, as illustrated in this conversation with a sixth grader.

Researcher: Are there any numbers between 0 and 1?

Child: No.

Researcher: How about one-half?

Child: Yes.

Researcher: About how many numbers are there between 0 and 1?

Child: Wait a minute. There's 1 / 2, 1 / 3, 1 / 4, 1-over all the way up to 10.

Researcher: Suppose you divided 2 in half and got 1 and then divided that number in half. Could you keep dividing forever?

Child: No, after 1 is zero. Zero is nothing else. If we kept dividing 1 / 2, [we'd get] 1 / 1, then 0 / 1, and 0 / 0, and that's it.

Researcher: Would you ever get to zero?

Child: Yes, zero is the last number.

Three additional years of math education did not help this sixth grader understand the relation between fractions and division any better than the third grader quoted earlier. This relation is critical to conceiving of rational numbers as distinct from whole numbers.

When children are asked, "Why are there two numbers in a fraction?" those who answer in terms of division—that the top number is *divided by* the bottom—demonstrate several other hallmarks of understanding rational numbers. They can order fractions like 1 / 75 and 1 / 56 correctly, they can order decimals like 0.65 and 0.8 correctly, they spontaneously acknowledge the existence of numbers between 0 and 1, and they judge that numbers can be divided indefinitely without ever reaching zero.[36]

Division paves a new path in the landscape of possible numbers, but research in math education suggests that division is best learned outside the context of numbers—as the partitioning of continuous quantities rather than the combination of discrete integers. Elementary schoolers have no trouble matching amounts based on proportion, such as matching a thirty-six-inch yardstick covered two-thirds in paint with a twelve-inch ruler covered two-thirds in paint. Even though a yardstick is three times longer than a ruler, children ignore the size difference and focus on the relative amount of paint. But if the yardstick and ruler are broken into units, children match them by units instead.[37] The units trigger the well-practiced routine of counting, and counting triggers the well-understood concept of whole numbers.

Still, counting can be subverted if children are encouraged to focus on proportions and given sufficient practice at doing so.[38] A child who practices matching yardsticks to rulers based on the proportion covered in paint will continue matching by proportions even when the painted areas are broken into units that do not match in number.

Focusing children's attention on proportions has proven successful at teaching them about rational numbers in general.[39] The standard approach to teaching rational numbers begins with fractions, illustrated as discrete pieces of a predivided pie, and then transitions to decimals, followed by percentages. But the opposite is more effective. Children have an intuitive grasp of percentages, which they encounter when working with measuring tools like tape measures and measuring cups. Children have no trouble distinguishing a measuring cup that is half full from one that is a quarter full or a tenth full. They can learn to associate these differences with percentages (50 percent, 25 percent, 10 percent) and then learn how to write these percentages as decimals (0.50, 0.25, 0.10) or fractions (50 / 100, 25 / 100, 10 / 100). This last step introduces the concept of division—division by 100—but only after children have a firm grasp on what's being divided and how.

Mapping quantities to notations and notations to operations is not easy. But this task is made easier by grounding children's understanding in quantities they can directly perceive and notations they can directly interchange. Children's first foray into rational numbers is typically the unit fractions of 1 / 2, 1 / 3, and 1 / 4, but it's far from obvious what these numbers represent or how they are related. Even adults who are well-versed in all forms of rational-number notation prefer to represent proportions as percentages, not fractions.[40]

Rational numbers are a tipping point in mathematics education. Students who fail to grasp them fall behind when learning subsequent topics, such as algebra and trigonometry. Indeed, elementary schoolers' understanding of fractions predicts their math achievement several years later, as high schoolers.[41] This prediction holds even after controlling for knowledge of arithmetic, general intelligence, household income, and parents' level of education. The key factor is students' understanding of fractions as magnitudes. Students' ability to add and subtract fractions is much less predictive of later math achievement than their ability to compare and order fractions.[42] Students have to conceive of rational numbers as numbers, on the same number line as whole numbers, before they can manipulate them with confidence—and competence—in the context of an equation.

A Sense of Direction

When adults are asked to identify their favorite subject in math, most pick geometry.[43] Geometry seems more intuitive than arithmetic and algebra, perhaps because it is grounded in truths we can see. Socrates famously argued that knowledge of geometry is innate and that discovering geometrical principles is a matter of exploring intuitions we already hold.[44] He illustrated this argument with a conversation between himself and an uneducated slave, chronicled by Plato in the *Meno*. Socrates draws a square with sides measuring two units in length and asks the slave how long the sides would be if the square were double in area. The slave answers four, but Socrates shows him that a square with sides that long would be four times as large as the original square, not twice as large.

Socrates then guides the slave to see that the new square must have sides equal to the diagonal of the original square. Cutting the original square by its diagonal would yield triangles with half the square's area, and putting four of those triangles together, right angle to right angle, would create a square twice as large as the original. The slave was able to appreciate the logic of this argument despite his lack of formal education, which Socrates took as evidence that geometric knowledge is a human birthright.

The slave's discovery is impressive, but is it typical? Do people in general have access to such insights? In a recent study, researchers asked people of varying ages, from twelve to sixty-four, the same fifty questions that Socrates asked the slave.[45] They found almost perfect agreement between the responses of this sample and that of the slave. Surprisingly, though, half failed to learn the lesson the dialogue was intended to teach. When presented with a new square and asked to double its area, they embarked on nonstarter solutions, like doubling its sides, rather than reproducing the method of diagonal bisection modeled for them minutes earlier.

This result highlights a tension in spatial imagination similar to the tensions in numerical imagination described earlier: while some geometric ideas are grounded in perception and intuition, others require the discovery of novel operations.[46] This tension is well illustrated by a pattern-matching task devised by the psychol-

Figure 6.2 Which figure in each panel is the oddball? We have little difficulty identifying oddballs in the top panels, exemplifying metric or angular relationships (bottom-center, top-right, bottom-right) but have considerable difficulty doing so in the bottom panels, exemplifying directional relationships (bottom-left, top-left, bottom-right). (Adapted from Dehaene et al. 2006) (Courtesy of the author)

ogist Stanislaus Dehaene and his colleagues.[47] In this task, participants see six visual patterns and must identify the pattern that violates a geometric principle illustrated by the other five. They might see six sets of intersecting lines, where five intersect at right angles and one intersects at an oblique angle, or six triangles, where five are equilateral and one is isosceles.

Using this task, researchers have discovered remarkable consistency in the geometric principles we are attuned to and those we are not. Most people are attuned to principles defined by topology, angles, or distance, such as whether a line is curved, whether two lines meet at right angles, or whether the sides of a triangle have equal lengths. But most people are *not* attuned to principles defined by direction, such as whether two lines intersect to form an *L* or a backward *L*, whether a small figure is to the left or right of a larger figure, and whether two irregular shapes are duplicates or mirror images. In other words, we easily perceive differences in the size of an angle, the length of a line, and the curvature of a surface, but we are oblivious to differences in left-right orientation.

Figure 6.3 What's wrong with these images? Both have been mirror reversed, but our insensitivity to direction, as a geometric property, makes this change difficult to detect. (Wikimedia Commons)

This finding holds for children as well as adults, and it holds for people in nonindustrialized societies, like the Munduruku, as well as people in industrialized societies, like France.[48] It holds even for people who were born blind, if the task is administered using tactile displays.[49] Humans appear to have an innate sense of space structured by topology, angles, and distances. This sense emerges early in development, regardless of where we grow up and whether we perceive space with our eyes or only our bodies. But it has a profound limitation: it makes no distinction between left and right—a property known as "chirality."

At first blush, chirality doesn't seem very important. Does it matter that we might confuse left-handed scissors with right-handed ones? Or whether the salad fork goes to the left of the plate or the right? Probably not. But many other aspects of modern life depend entirely on chiral mechanisms and chiral symbols. Screws tighten to the right but not the left. Cars drive on the right but not the left (in most countries). Maps are worthless if held upside down, inverting their directions. Drugs can become deadly if their chemical structure is inverted. The letter *b* becomes *d* if written in reverse, and nearly all letters become illegible if reflected in a mirror.

Chirality is an underappreciated aspect of many modern systems. We forget how we had to master chirality when first engaged with those systems, especially the writing system. Literate adults write messages in one direction and only one direction, but children who are just learning to write will scrawl letters in both directions. In fact, when five-year-olds are asked to write their name next to a dot placed at the far right of a sheet of paper, they typically start at the dot and write their name backward rather starting a few inches from the dot and writing their name forward.[50]

Training ourselves to write in a single direction requires overcoming a deep-seated bias to lump similar shapes into the same category. A rose is still a rose if flipped from left to right, but a *b* becomes a *d* and a *p* becomes a *q*. Studies that have looked at the neurons involved in shape recognition find that these neurons are sensitive to viewpoint but not chirality. A neuron that fires when you look at a right angle will fire regardless of whether that angle points to the left or to the right.[51]

Our insensitivity to chirality suggests that it's not a critical feature of how we interact with natural environments, but it has become a critical feature in how we interact with modern artifacts, from screws to letters to maps. Such artifacts embody spatial distinctions that transcend the innate organization of spatial imagination.

Maps, for instance, embody directional information that is opaque on first encounter. Children have little trouble using maps to identify distinct landmarks, but they routinely confuse locations differentiated only by direction.[52] If shown a map with three locations arrayed in an isosceles triangle, children are prone to confuse the two locations at the base of the triangle. They can use the map to navigate to the top location because this location differs from the other two by distance, but they will vacillate between the two locations at the bottom, navigating to the correct location only half the time.

As opaque as chiral tools may first appear, learning to use these tools expands our spatial imagination. Children who are better at reading maps are also better at identifying direction-based patterns in the pattern-matching task described previously, such as identifying the one angle that points in a different direction than the others.[53] Children who are better at reading words are also better at writing them, printing their letters in the correct direction and using the correct form of mirror-image letters like *b* and *d*.[54] Humans have created many tools that help us keep track of direction, such as maps, or that use direction as a cue to other information, such as letters; the more we use these tools, the more we attend to direction as a distinct feature of space.

The mutual support between mathematical tools and mathematical imagination can also be seen in the domain of arithmetic. Humans have created many tools for keeping track of quantities and calculations, including counting boards, ledgers, spreadsheets, and slide rules, and experience with these tools enhances our ability to perform arithmetic operations.

Consider the abacus. This tool allows users to create and manipulate a visual representation of multidigit numbers, using beads to represent quantity and columns to represent place value. People who are well-practiced at using an abacus no longer need a physical abacus to take advantage of its affordances: they can visualize an abacus in their mind and then mentally simulate moving its beads. Mental-abacus users can perform multidigit addition problems quickly and efficiently, and this ability improves with practice.[55]

The very best mental-abacus users can sum ten ten-digit numbers in their head in a matter of minutes.

Tools like the abacus and the map provide an efficient means of applying the abstract operations inherent in mathematics. Such operations allow us to contemplate novel mathematical possibilities, but they are prone to misuse and misapplication, such as when we make calculation errors that defy commonsense understanding of the quantities being compared or use a formula that defies commonsense understanding of the situation being modeled. Grounding abstract operations in concrete procedures helps us avoid these problems by tying them back to the physical manipulations they are meant to represent. Striking the right balance between the concrete and the abstract is key to expanding mathematical imagination in a sustainable way, but striking this balance can be as difficult to achieve as discovering the relevant abstractions.

Take, for example, the proof we discussed at the start of the chapter on the relation between odd numbers and square numbers. Imagining square numbers as actual squares feels like a cheat—a superficial trick for appreciating a deep insight that only mathematicians can grasp. But the connection between square numbers and geometric squares is as genuine as the connection between counting and adding. And the discovery of that connection is as innovative as the discovery of the formal proof that consecutive odd numbers sum to square numbers. If the formal proof requires years of advanced mathematics to understand, then the visual proof is surely more impactful because it makes this insight available to everyone. Forging connections between reality and idealized versions of reality is itself a feat of imagination.

7 Ethics

Expanding Our Moral Imagination

When Benjamin Franklin was a young man, he decided he needed to get his moral life in order.[1] He noticed troubling lapses in conduct that he hoped to correct through reflection and self-examination. Franklin compiled a list of thirteen virtues that he vowed to follow more regularly over the course of a year: temperance, silence, order, resolution, frugality, industry, sincerity, justice, moderation, cleanliness, tranquility, chastity, and humility. He then created a grid to keep track of his moral behavior. Day by day, Franklin monitored his conduct for lapses in each virtue, marking his grid with every lapse. His goal was to make fewer marks with each week, but the task was onerous. The running record of moral lapses opened Franklin's eyes to the complications of moral life. "I was surprised," he wrote, "to find myself so much fuller of faults than I had imagined."

Moral life is complicated not because we are unaware of the virtues we should uphold but because we should uphold so many. Franklin identified thirteen, but his list is far from complete. Consider

Examination of Virtues

Eat not to dullness;
Drink not to elevation.

	Sun.	M.	T.	W.	Th.	F.	S.
Tem.							
Sil.							
Ord.							
Res.							
Fru.							
Ind.							
Sinc.							
Jus.							
Mod.							
Clea.							
Tran.							
Chas.							
Hum.							

Figure 7.1 Benjamin Franklin attempted to organize his moral life by keeping a daily log of his adherence to thirteen virtues: temperance, silence, order, resolution, frugality, industry, sincerity, justice, moderation, cleanliness, tranquility, chastity, and humility. (Wikimedia Commons)

the pledge that boy scouts make to be trustworthy, loyal, helpful, friendly, courteous, kind, obedient, cheerful, thrifty, brave, clean, and reverent. Boy scouts agree with Franklin that cleanliness and thriftiness are virtues (even if they might pursue them less successfully than Franklin), but those are the only points of commonality. Boy scouts pledge themselves to several distinct virtues, all of which can be as difficult to uphold as those identified by Franklin.

Part of the difficulty is maintaining the required vigilance and fortitude, but a greater difficulty is figuring out how to coordinate one virtue with another. Moral values conflict. Kindness sometimes requires dishonesty; helpfulness sometimes requires discourtesy; bravery sometimes requires disobedience. Even a single moral value can conflict with itself. Loyalty to one friend can constitute disloyalty to another; thriftiness in one situation can lead to wastefulness in others; reverence for one's own beliefs can foster irreverence for others'.

How do we cope with such conflict? Emotions are one solution.[2] Rather than systematically map the moral landscape, as Franklin attempted to do, we usually let our emotions be our guide. Emotions like empathy, compassion, guilt, shame, and disgust are tailor-made for moral situations, and they do a fair job of steering us toward virtue and away from vice. But, as many psychologists have noted, emotion-based decisions can be as disjointed as the values they help us navigate. Empathy drives us to assist hurricane victims and cancer patients but allows us to ignore the chronic suffering of the poor.[3] Disgust drives us to condemn pedophiles and rapists but can fuel hostility toward individuals with different sexual orientations or gender identities.[4]

A different solution to the problem of organizing moral life are moral principles, or commitments to ethical outcomes that transcend the pull of momentary emotions and the ambiguity of competing virtues. They provide a check on impulses that might lead us astray from our long-term goals or preferred modes of conduct, and they highlight courses of action that we might not otherwise imagine, let alone undertake.

Moral principles help us think through the permissions and obligations of everyday life, as well as consider new moral possibilities altogether, such as caring for the welfare of strangers, upholding the dignity of the disabled, and improving the well-being of nonhuman animals. Moral principles move us beyond local concerns—about ourselves, our family, and our community—to universal ones. When we view access to education as a moral right, we believe it applies to all children, not just our own. When we view fair compensation as a moral obligation, we believe it applies to all employees, not just ourselves.

Moral principles, like causal principles and mathematical principles, must be learned. They are rarely intuited from daily life because they deviate from our earliest social inclinations. Young children are not moral angels, as commonly believed—they are moral pariahs.[5] Across many studies, they have shown themselves disinclined to share with others, disinclined to help others, and disinclined to interact with people outside their social group. In fact, they actively dislike people outside their social group, preferring to see them treated poorly rather than well.

Our moral imagination is initially characterized by parochiality and self-interest and may remain so if not for moral instruction. Commitment to universal equality and universal rights is a foreign state of mind, both historically and developmentally.[6] In the following sections, we will explore the development of three principled distinctions in moral imagination: that bad outcomes are not always caused by wrongdoing, that the way things are is not always the way they ought to be, and that an equal distribution of resources is not always an equitable distribution.

These distinctions are a far cry from the higher-order principles that define modern ethics, such as Immanuel Kant's categorical imperative (that we should treat moral duties as universally applicable),

Jeremy Bentham's utilitarian calculus (that we should maximize pleasure and minimize pain), or John Rawls's veil of ignorance (that we should strive for outcomes that all segments of society would endorse), but they are prerequisites for learning such principles. Children must appreciate intentional harm before they can appreciate the imperative to avoid it, and they must appreciate inequity before they can appreciate its structural causes. Learning a new moral principle opens our mind to new moral possibilities, and contemplating such possibilities can, in turn, open our minds to new principles.

Distinguishing Wrong from Bad

Of all the moral concerns that might occupy our mind, harm is chief among them. When psychologists survey people's everyday moral lives, by sending them random texts and asking them to report on the moral activities they may have experienced or witnessed within the last hour, the majority of activities that people report involve harm—either causing it or alleviating it.[7] Benjamin Franklin may have monitored his conduct for lapses in moderation, humility, and frugality, but these lofty virtues barely scratch the surface of everyday morality. We are not deeply concerned when the people around us eat too much, boast too much, or spend too much money, but we are deeply concerned when they hit, cheat, or steal. The same is true for ourselves when monitoring our own conduct.

A better guide to everyday morality than Franklin's list of virtues are the ten commandments, which specifically forbid harm: do not kill, do not steal, do not cheat, do not lie (to paraphrase a few). But avoiding harm is harder than it sounds. Harms come in many flavors, as well as many forms. Some are planned, whereas others are merely foreseen. Some are caused by action, whereas others are caused by inaction. They can be immediate or distant, direct or indirect, malicious or instrumental. Harms are generally associated with wrongdoing, but not all harms are wrong, and not all wrongs are harmful.

Consider the most egregious harm: murder. Most legal systems do not treat murder as uniformly wrong but contingently wrong—contingent on the motives and circumstances of those involved. US law, for instance, distinguishes between first-degree murder, or killing intentionally and without provocation; second-degree murder, or killing with the intent to harm rather than kill; voluntary manslaughter, or killing in the heat of passion; involuntary manslaughter, or killing by recklessness or negligence; and self-defense, or killing in order not to be killed. These distinctions are nuanced but generally comprehensible to the average adult.[8] We can parse the intentions behind a harm as distinct from the harm itself, allowing for the possibility that a person could be responsible for murder but not culpable. As the murderesses in the musical *Chicago* explain, "it was murder but not a crime."

The idea of a crimeless murder stretches imagination. It lies outside the range of moral possibilities we typically consider, as well as those accessible to young children. Unlike adults, young children do not parse the intentions behind a harm from the harm itself. Their conflation of harm with wrongdoing was first documented by the psychologist Jean Piaget, who asked children whether it's worse to break fifteen glasses by accidentally knocking over a tray or break a single glass while purposefully rummaging through a cupboard. He found that children typically judged the first scenario as worse, privileging the severity of the outcome over the intent of the actor.[9]

Piaget's findings have been replicated by other researchers, who find that children's earliest evaluations of wrongdoing are based on outcomes rather than intentions.[10] In one study, preschoolers were told stories about two types of actors: an actor who causes harm by accident and an actor who attempts to cause harm but fails. They were then asked to decide who is naughtier and who should be punished.[11] For instance, one pair of stories contrasted a boy who trips on a rock and accidentally pushes someone over with a boy who attempts to push someone over but trips on a rock and misses. Adults judge the second boy as naughtier and more deserving of punishment, but preschoolers are ambivalent in their judgments, sometimes selecting the first boy and sometimes selecting the second. They fail to see the actors' intentions as consistently more important than the outcomes they bring about.

This failure is not absolute. Preschoolers recognize that attempted harms are wrong; they just don't think that attempted harms are more wrong than accidental harms. Preschoolers also recognize that actors who attempt harm deserve punishment, but they don't think those actors deserve a lot more punishment than actors who cause harm accidentally. If preschoolers wrote the laws on murder, they wouldn't assign more culpability to attempted murder than involuntary manslaughter.

A central reason preschoolers fail to distinguish harmful intentions from harmful outcomes is that they are still developing a "theory of mind," or the ability to analyze others' behavior in terms of their underlying mental states. A classic measure of theory of mind is the false-belief task, where children are asked to predict the behavior of a character whose beliefs conflict with reality. They might be told, for instance, that a girl was using markers at an art table before she went outside for recess and that, while she was at recess, her teacher moved the markers to a cabinet. They would then be asked where the girl will look for the markers when she returns: the art table or the cabinet. Children over the age of four typically say the girl will look for the markers at the art table, where she left them, but younger children say she will look for the markers in the cabinet, where they actually are.

Children who pass this task, acknowledging that people sometimes act on the basis of false beliefs, are more forgiving of actors who cause accidental harms.[12] If told of an actor who threw away someone else's possessions by accident, they rate this actor as less naughty and less deserving of punishment than do children who fail the false-belief task. An appreciation of false beliefs allows for an appreciation of how someone with good intentions could mistakenly cause a harmful outcome. Consistent with these findings, adults with autism also fail to see the excusability of accidental harms. Autism impairs a person's ability to simulate others' beliefs, and, accordingly, adults with autism judge accidental harms to be as wrong as attempted harms.[13]

Distinguishing accidental harms from attempted harms has proven to be a particularly sensitive test of mature moral evaluation because it requires us to coordinate outcomes with intentions. Neuroscientists have investigated where in the brain this coordination takes place and have identified an area on the right side of the brain, at the junction of the temporal and parietal lobes. This area, the RTPJ (right tempoparietal junction), is particularly active when we reason about attempted harms or harmful intentions in the absence of harmful outcomes.[14] In fact, when people are asked to evaluate the wrongfulness of an attempted murder, their judgments become more lenient if this area is impaired. By applying a magnet to people's skulls at the site of the RTPJ, researchers have induced them to view attempted murder as more permissible.

Disrupting the RTPJ does not induce leniency toward intentional harms, though. This area helps us recognize a unique kind of moral offense, where wrongdoing can be identified independent of harm. Only someone who has embraced the principled distinction between harmful intentions and harmful outcomes could see the mere attempt to cause harm as falling within the realm of moral concern. Likewise, only someone who has embraced this distinction could see a harm that was not intended as morally forgivable.

That said, the distinction between accidental harms and attempted harms does not come for free with a mature theory of mind; it must be modeled and practiced. In some cultures, like Fiji, reasoning about others' mental states is discouraged. Fijians subscribe to the philosophy that no one can truly know what another person is thinking, and, accordingly, they judge attempted harms to be less wrong than accidental harms. Fijians recognize that actions are undergirded by intentions, but they do not speculate on those intentions, nor do they regard an intention in the absence of an outcome as morally meaningful.[15]

Unlike Fijians, Westerners routinely privilege intentions over outcomes when making moral judgments, but even Westerners will shift their focus from intentions to outcomes if they are distracted. When Westerners make moral judgments while also performing a second task, such as listening to an audio recording and repeating it out loud, they stop judging accidental harms as acceptable and begin judging them as wrong, sometimes even as wrong as attempted harms.[16]

In short, accidental harms and attempted harms become distinct moral possibilities only after children have learned that intentions weigh more than outcomes. And only when this principle is culturally endorsed. And only when it is actively deployed. Learning to distinguish harm and wrongdoing can expand our moral imagination, but doing so requires cognitive effort and cultural support.

Distinguishing Ought from Is

Intentional harms are actions we are obliged to avoid. On the other side of the moral spectrum are actions we are obliged to perform: telling the truth, helping our friends, caring for our children. How do we know what actions are morally prescribed? Philosophers have struggled with this question for centuries, but children answer it quite easily: anything that's typical. If people *do* perform an action, then they *should* perform that action. Is implies *ought* in young children's minds.

A salient example of this conflation comes from a study by psychologist Marco Schmidt.[17] Schmidt and his colleagues showed three-year-old children novel actions performed with novel tools, such as pushing a marble across a table using a T-shaped rod. Children were then given the tool and allowed to play with it themselves. Later, a second researcher entered the room, picked up the tool, and performed the same action in a different way, such as gripping the T-shaped rod backward and pushing the marble with its handle.

Children did not watch the second researcher passively. Instead, they protested her actions, letting her know she was "doing it wrong" and showing her the correct way to do it. Children protested the second researcher's actions regardless of whether the tool was a rod, hook, or slider and regardless of whether the action she performed was pushing, pulling, or tapping. They protested her actions when the first researcher explicitly told them how to use the tool, as well as when they observed the first researcher from afar, without receiving any explicit instruction. The consistency of children's protests inspired Marco and his colleagues to label them "promiscuous normativists": they promiscuously infer norms from a single observation.

Promiscuous normativity seems benign when it comes to tool use but can be pernicious when it comes to cultural norms—norms about what clothes to wear, what games to play, or what foods to eat. Many Germans eat schnitzel, and many Koreans eat kimchi, but does that mean Germans ought to eat schnitzel and Koreans ought to eat kimchi? Is it wrong for a German to eat kimchi or a Korean to eat schnitzel?

Young children seem to think so. When introduced to novel social groups, they claim that it is wrong for a member of one group to perform the customs of the other. If told, for instance, that Hibbles eat purple berries and Glerks eat orange berries, preschoolers claim that it is wrong for a Hibble to eat an orange berry and wrong for a Glerk to eat a purple berry, judging those who eat a different kind of berry as "very, very bad."[18]

Children's condemnation of atypical social behaviors are as consistent as their condemnation of atypical tool use. They say it's wrong for a Hibble to eat orange berries if they hear an adult proclaim "Hibbles eat purple berries" or if they simply watch Hibbles eat purple berries.[19] They say it's wrong for a Hibble to eat orange berries if Hibbles are humans or nonhuman animals.[20] And they say it's wrong for a Hibble to eat orange berries regardless of whether they live in a society that emphasizes individual freedoms like the United States or a society that emphasizes conformity to social norms like China, though judgments of wrongness are stronger and longer-lasting in the latter.[21]

Children's conflation of is and ought may stem from the belief that what's actual is ideal—that we live in the best of all possible worlds. This idea, famously satirized by Dr. Pangloss in Voltaire's *Candide,* connects is with ought through the notion of goodness: the world is good, and good things ought to be.

Children do, in fact, endorse this logic. They claim that typical behaviors ought to be performed because typical behaviors are good. For instance, children claim that people ought to give flowers on Valentine's Day because flowers are the most beautiful gift and that people ought to drink orange juice at breakfast because orange juice is the most invigorating juice.[22] Children sometimes disagree on the particular reason that flowers are the best Valentine's Day gift or orange juice is the best breakfast juice, but they agree in general that something intrinsic to the situation makes it good and therefore right.

Another reason children conflate is with ought is that differentiating them requires thinking of alternatives, and thinking of alternatives requires a degree of reflection that many children lack. As we saw in Chapter 2, children easily jump from the perception that something is unusual to the conviction that it's impossible. Further reflection is required to differentiate events that could occur under different circumstances from those that could never occur. Children likely make the same mistake when they jump from the perception that something is typical to the conviction that it's prescribed. Indeed, asking children to slow down and think before passing a moral judgment reduces the likelihood that they will say an atypical behavior is wrong, at least for older children.[23] Preschoolers judge atypical behaviors as wrong regardless, possibly because they have trouble identifying circumstances under which such behaviors might occur.[24]

These findings suggest that distinguishing ought from is entails two challenges: the challenge of recognizing that typical behavior is not intrinsically good and the challenge of identifying circumstances that would lead a person to deviate from such behavior. We must learn to appreciate the principled distinction between convention and obligation but also to apply that distinction in the moment. These challenges are most salient for children, who are just learning the norms that guide social behavior, but they remain salient even for adults.

As adults, our perception of what's right or wrong continues to fluctuate with our observation of what's typical or atypical, despite a general appreciation that these notions are distinct. Donating money to a common cause, like repairing a road, is viewed as commendable, whereas keeping one's money and letting others repair the road is viewed as reproachable, but these views fluctuate with our observation of how many other people are donating. The fewer people we observe donate, the less wrong we judge their selfishness and the less we attempt to punish it.[25]

In fact, global studies of moral behavior find that people are more likely to engage in immoral behavior if they live in a country where those behaviors are commonly observed. People who live in countries where tax evasion, corporate malfeasance, and political corruption are common, like Guatemala and Tanzania, are more likely to cheat in an anonymous dice game. On the flipside, people who live in countries where cheating is rarer, like Austria and the United Kingdom, rarely cheat themselves.[26] Moral imagination is shaped by the moral environment. Behaviors that are normal in our environment become imbued with moral acceptability or even approbation.

That said, our perception of what's normal is shaped not just by what we see but also by what we want to see. We want our friends and neighbors to tell the truth and contribute to the public good, but we see them lying and acting selfishly on many occasions, so we deem normal behavior as falling somewhere in between. What, for instance, would you estimate is the normal number of lies a person tells each week? Studies reveal that our estimates fall between the average and the ideal. We think it's normal to tell eight lies per week, yet we also think people tell twenty-four lies per week on average but should only tell three lies per week in an ideal world. Likewise, our estimate of the percent of students who normally cheat on an exam (16 percent) falls between our estimate of the average percent (35) and the ideal percent (4).[27]

We recognize that average behavior differs from ideal behavior, yet we maintain a third notion of behavior—"normal"—that conflates the two. And it's this third, conflated notion that comes to mind most easily. If asked how many lies people tell per week, without any qualification of what's normal, average, or ideal, we respond with what we think is normal: eight.[28] What we imagine to be true is influenced by what we believe ought to be true.

Another quirk of the is–ought divide that remains challenging even for adults is imagining how events might unfold differently from how they ought to unfold. We anticipate that people will conform to social norms, and when they do not, it takes additional effort to process the violation. Consider this story about sexual temptation: "Valerie had been married to Daniel for over a year. Unfortunately, due to his job, she saw very little of him during the week. Recently she noticed how attractive Daniel's best friend was. The friend occasionally came to her house to use Daniel's computer. Valerie had all the opportunity she needed to seduce him while Daniel was at work." How might the story end? How *should* the story end?

Most people think the story should end with Valerie fighting her temptation to cheat. And if we learn that Valerie succumbs to her temptation instead, it takes us longer to process that ending than the ending we expect.[29] Similar results have been observed when people are asked to make moral decisions themselves in a multiple-choice format. We spend little time processing choices we deem immoral, and we report that it didn't take much effort to avoid them.[30] Immoral choices are not viewed as choices at all.

Our tendency to focus on prescribed behavior is surely adaptive; we waste little time contemplating behaviors that are not prescribed and thus rarely come to pass. But our sense of what people ought to do is a poor guide for deciding what they can do or what they should be allowed to do. Moral progress depends on making a principled distinction between what's normal and what's right. Many behaviors that were once common are now considered immoral, and many behaviors that were once considered immoral are now common.

Consider marriage conventions. For much of human history, marriage between older men and adolescent girls was common, whereas marriage between individuals of the same sex was forbidden. Now the former is forbidden, and the latter is common, at least in Western countries. The moral possibility of same-sex marriage and the moral turpitude of underage marriage became appreciable only when enough people were able to separate what marriage was from what marriage ought to be. Moral imagination cleaves closely to social reality, but we can separate them with reflection on the immoral aspects of everyday behavior, as well as the moral standing of behaviors we rarely (or never) perform.

Distinguishing Fair from Equal

In a popular meme about fairness, three children stand behind a fence trying to watch a soccer game on the other side. One is tall, one is average height, and one is short, but none are tall enough to see over the fence. In the next panel, all three children are standing on boxes. The tallest child can now see the game, but the other two cannot. This panel is labeled "equality." In the third panel, the children are once again standing on boxes, but the average-height child is standing on two boxes, stacked vertically, and the short child is standing on three. Now all three children can see the game. This panel is labeled "equity."

The point of the meme is that an equal distribution of resources is not always a fair distribution. Sometimes the recipients need to be taken into account, not just the resources being distributed. But young children have trouble coordinating the two in their moral imagination. They are dutifully committed to distributing resources equally but are not particularly concerned about achieving equity.

A striking demonstration of children's early commitment to equality comes from a study where preschoolers were asked to distribute resources that could not be distributed evenly, such as distributing five erasers to two other children. When children had allocated two erasers to each recipient, they were then faced with a conundrum: should they give the fifth eraser to one of the two recipients, creating unequal distributions, or should they throw it away? Almost all children opted to throw it away. They opted to throw away the extra resource regardless of whether it was an eraser or a more valuable resource like a candy bar. And they did so regardless of whether they came from high-income households or low-income ones, where resources like candy bars are harder to come by.[31]

Children's insistence that resources should be distributed equally emerges early in development. One-year-old infants are surprised

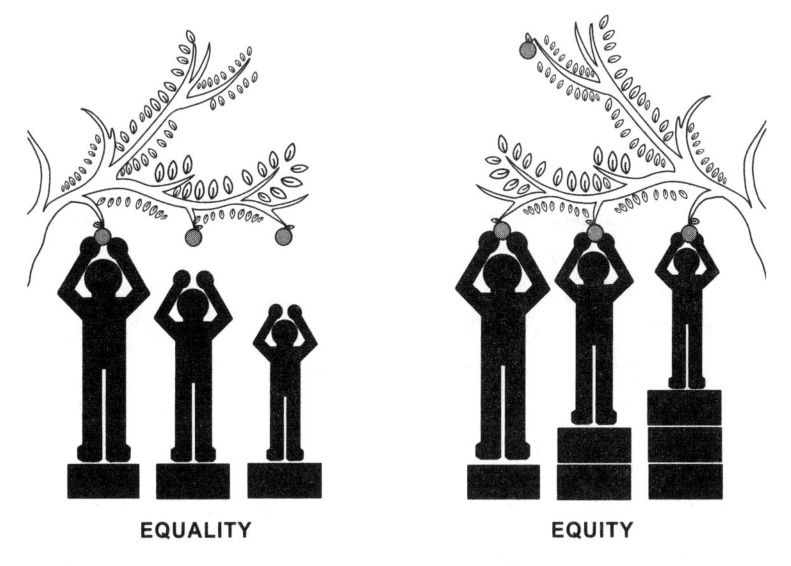

Figure 7.2 Children are strongly committed to equality but must learn that an equal distribution of resources (left) is not always a fair distribution (right). (Courtesy of the author)

when they watch someone with two toys distribute them unequally, giving both toys to one recipient and none to the other. They stare longer at this event than when they watch the toys being distributed equally, one to each recipient.[32]

By age three, children actively protest unequal distributions.[33] They let the distributor know that there's a discrepancy in their allocations and request that they be fixed. Sometimes they even express disdain, declaring "That's not fair!" or "That's not right!" Three-year-olds get most upset when they receive fewer resources than someone else, but they also get upset when they witness others receive unequal shares in distributions that do not involve themselves.

This fixation on equality would appear to be a moral virtue, but it can lead children to ignore circumstances where equal distributions are not warranted, either because the recipients are not equally deserving of the resource or because they do not have equal need of it. Consider cases of collective effort, like group projects, where some people in the group exerted more effort than others. Should the spoils of these projects be distributed equally?

Preschoolers recognize that group members who work harder deserve more resources, but they only give them more resources when forced to do so—if, for instance, they must distribute five cookies between two recipients without throwing any away. If they can distribute resources equally, they will, ignoring differences in effort. Older children are more likely to take effort into account, but their rewards to the more deserving recipient are nominal—three cookies out of five, for instance, even when the more deserving recipient did all the work.[34]

Young children think resources should be distributed equally even when recipients differ in need. In one study, preschoolers were asked to allocate stickers to two recipients—one who already had several stickers and one who had none. They tended to give both recipients the same number of stickers. If they watched someone else distribute stickers, they rated the distributor as "very good" when she distributed stickers equally but rated her as only moderately good—or even bad—when she gave more stickers to the puppet who had none.[35]

Preschoolers also think it's bad to redistribute wealth, from the rich to the poor. When told about a character who steals from the rich to give to the poor, à la Robin Hood, they judge that character's behavior as wrong. In fact, they think it's as wrong to steal from the rich to give to the poor as it is to steal from the poor to give to the rich.[36] Stealing, from their perspective, is just wrong, regardless of who is stolen from and for what reasons.

In short, considerations like need and merit do not loom large in children's evaluation of how resources should be distributed. This finding has been documented in societies around the globe, from China to Chile and Jordan to Norway.[37] The belief that resources should be divided equally is seemingly universal among preschoolers. It does not shift until around age six, possibly as a result of schooling. School is a context where children are introduced to norms that transcend the parochial concerns of their own family. "One for me, one for you" is a workable strategy when dividing resources among siblings or cousins but will not necessarily work when a wider variety of people and interests are at stake.

Other norms that come into play are norms about compassion, partiality, and wastefulness. An even split is often the most compassionate, least partial, and least wasteful distribution of resources, but sometimes it's not. A distributor who deprives someone in need lacks compassion; a distributor who rewards someone who doesn't deserve rewards appears partial; and a distributor who leaves resources undistributed is wasteful. As children learn to appreciate such contingencies, they cease distributing resources equally and begin distributing them equitably.

This transition is bumpy, though. When children become willing to entertain the possibility of nonequal splits, they have trouble distinguishing legitimate reasons for favoring a recipient from illegitimate ones. Faced with the question of how to distribute six bananas to two recipients, six-year-olds will give more bananas to the recipient who is hungrier or to the recipient who did more chores, but they will also give more bananas to a recipient who simply declares "I want more!"[38]

It is particularly difficult for children to balance compassion and partiality when they themselves are one of the recipients.[39] Faced with the decision of how to divvy up two candies between themselves and another child, preschoolers almost always keep both. Their egalitarian stance toward others falls apart when they stand to benefit from the distribution. Older children are more willing to share candies with others but not if it puts them at a disadvantage. Given the choice of dividing candies equally ("one for me, one for you") and dividing candies in favor of someone else ("one for me, two for you"), children of all ages pick the first option. They prefer to see resources distributed equally than to see someone else receive more, even when the latter could be achieved at no cost to themselves.

And while children are deeply bothered by the possibility that someone might accrue more resources than they will, they are not bothered by the possibility of accruing more resources than someone else. Children of all ages reject unequal splits if the split would favor someone else ("one for me, four for you"), but they are willing to accept the same split if it would favor themselves ("four for me, one for you"). This kind of inequity—advantageous inequity—is generally viewed as acceptable across ages and cultures.[40] In some societies, like Canada and Uganda, children begin to reject advantageous inequity in middle childhood, but they never reject advantageous inequity as reliably as they reject disadvantageous inequity. Children are okay with inequity so long as it works to their advantage, as are many adults.[41]

Equity is thus far more complicated than equality. What counts as an equitable distribution changes with the recipients, the resources, and the situation, not to mention the interests of the distributor. There's no single strategy for divvying up resources that would always be identified as equitable. But equitable strategies are, in fact, attainable; they just require more imagination than the strategy we prefer as children: "one for me, one for you."

The first step toward embracing equity is seeing it as distinct, in principle, from equality. This distinction allows us—or may even compel us—to rethink our earlier moral inclinations. Rewarding a social loafer as generously as a hard worker turns from being fair to being indifferent, even partial. Throwing extra resources in the trash turns from being prudent to being wasteful. And redistributing money from the rich to the poor turns from an act of theft to an act of compassion. Notions of equity not only forge new paths in the landscape of moral possibility but also relabel familiar ones.

Prohibited paths sometimes become prescribed, and prescribed paths sometimes become prohibited.

Balancing Principles with Particulars

The tale of Robin Hood is a telling example of how moral principles can reshape the landscape of moral possibility. On its surface, the tale is about a serial larcenist who uses coercion and violence to take resources away from those who lawfully own them. And consistent with this interpretation, young children view Robin Hood's deeds as "not okay."[42] But older children and adults hail Robin Hood as a hero and see his thievery as an act of justice.

The reason we come to view Robin Hood as a hero has to do with the circumstances surrounding his actions. The kingdom of Nottingham is run by callous aristocrats who earned their money by corrupt means and spend it on luxury goods, depriving the kind people of Nottingham of basic necessities like food and shelter. But how might we view Robin Hood's actions in the context of a more nuanced social order? What if Nottingham's ruling class had earned its wealth through hard work and prudent investment? What if they regularly used that wealth to build and maintain the kingdom's infrastructure? What if the people of Nottingham were not uniformly poor but varied in wealth? What if Robin Hood's spoils covered only a portion of the people's needs, leaving many just as destitute?

Applying moral principles, like the principle of distributive justice, is rarely easy. It requires coordinating abstract ideals with the nuances of social situations, similar to the challenge of applying abstract causal mechanisms to concrete events (discussed in chapter 5) or applying abstract operations to concrete quantities (discussed in chapter 6).

To put a finer point on this predicament, consider the seemingly obvious principle that, in life-or-death situations, we should maximize the number of lives saved and minimize the number of lives lost. Fewer deaths are always better than more, right? But now consider the bleak fact that thousands of people are currently dying of organ disease or organ failure and need a new organ to stay alive. Some people need a new heart; others need a new kidney; still others need a new lung. A single healthy person could donate all such organs. They could donate one heart, two kidneys, and two lungs, saving five people's lives in exchange for their own. Should society allow people to make this sacrifice? Should society force some people to make it?

If you felt compelled by the principle that we should maximize the number of lives saved, you likely felt less compelled after considering the possibility of sacrificial organ transplants. This possibility is consistent with the principle but inconsistent with other commitments we hold, such as commitments to personal dignity, consent, and control. Sure enough, when people hear about sacrificial organ transplants, they change their mind about the value of maximizing lives saved. Their support for the statement "always take whatever means necessary to save the most lives" drops by a quarter.[43]

This interplay between moral principles and moral particulars occurs in the opposite direction as well. People who judge that sacrificial organ transplants are immoral can be swayed to accept such transplants if they are informed of the rationale behind them. The same holds for other emotionally aversive actions, such as smothering a crying infant to keep her cries from alerting enemy soldiers intent on killing you and your neighbors.[44] No one thinks that harvesting organs or smothering infants is a good idea, but we can be persuaded to accept the necessity of such actions in light of the principle that it's better to sacrifice one life to spare several others.

Moral principles shape our perception of moral particulars, and moral particulars shape our endorsement of moral principles.[45] Principles can nudge us toward actions we might normally avoid, and particulars can nudge us toward principles we might normally reject. The interplay between these considerations is a necessary tension in moral imagination.[46] It's what keeps us from hoarding resources and ignoring the needs of others while also keeping us from relinquishing all our resources and living only for others. It's what allows us to create and enforce universal laws while also exhibiting compassion for those who break the laws by accident or for just cause.

In this way, moral principles function more like guidelines than absolutes. They help us identify new moral possibilities—untapped permissions and unrecognized obligations—but their application is constrained by the particulars of social life. As Benjamin Franklin astutely observed, the pursuit of virtue and the avoidance of vice is "a task of more difficulty than imagined," even for those with a moral imagination as principled as Franklin's.

Part III

Expanding Imagination by Model

8 Pretense

Expanding Our Symbolic Imagination

In the spring of 2020, parents began noticing their children engage in a new form of pretend play: coronavirus games.[1] Siblings diagnosed each other with the disease. Preschoolers performed swab tests with Legos and plastic dishes. School children substituted a spiky "CoronaBall" for their regular dodgeball. Others changed the rules of tag to create a socially distanced version where players tagged each other's shadows rather than their bodies. Was this new form of pretend play a sign of stress and anxiety? Shouldn't children be occupying themselves with fantasy games rather than the grim details of a pandemic?

As disturbing as coronavirus games might be to an adult onlooker, they are not a departure from children's typical style of play. Children routinely incorporate elements of the real world into their pretend games, including elements of the adult world. They pretend to be chefs preparing meals, cops arresting criminals, or teachers lecturing pupils. Their pretense is filled with real tools, real goals, and real occupations. The content might vary, but the focus on reality does not.

Consider this collection of anecdotes from Jerome and Dorothy Singer, two psychologists who have spent decades observing children at play: "We observe these little creatures talking to thin air, treating a small blanket or a stick like a baby that needs cuddling, having a teddy bear talk on a toy telephone in phrases suspiciously like a mother's. Sometimes we see a quite little boy carrying on bloody and bitter battles punctuated by imitative cannon noises, diving airplanes, with only the help of some props of wooden blocks and perhaps plastic toy soldiers. A three-year-old girl goes on at great length instructing her invisible friend Mrs. Puffum to be more careful—if she keeps on stepping in puddles she'll ruin her new white shoes."[2]

These anecdotes are intended to convey the variety of children's pretend play, but what might be most striking about them is their realism. It takes a fair amount of imagination to pretend that a stick is a baby or that a teddy bear can talk but much less to emulate a father cuddling a baby or a mother talking on the phone. It takes a fair amount of imagination to converse with an invisible friend but much less to converse about puddles and shoes. And while firing cannons and flying airplanes may be novel actions to a child, they are not conceptually novel—not something invented by the child.

Children's ability to disengage from a real situation in order to engage with a hypothetical one is, without doubt, a feat of imagination. But the content of that activity is not particularly imaginative. Children disengage from reality only to contemplate real-world activities.[3] The events they make up embody the same content and logic as those they observe. The same is true for adults. We create models of reality, in the form of stories and games, for the express purpose of exploring those models—testing their parameters and drawing out their implications. But we populate these models with real-world entities, and our exploration is constrained by real-world expectations. Creating models that deviate from ordinary experience takes genuine effort, as does exploring those models in ways that challenge, rather than reinforce, what we already know.

Many people believe that pretense starts out wild and unconstrained and only later turns into something structured and practical. Adults may daydream about what to say in a meeting or what to cook for dinner, but children are believed to immerse themselves in exotic scenarios, like building rocket ships or trapping velociraptors.[4] This belief turns out to be mistaken. Studies of pretend play find that pretense in childhood is highly realistic, thoroughly structured, and tediously rule-driven. Rocket ship building and velociraptor hunting are exceptions, not the rule. The pretend scenarios that children typically construct are much more mundane, like pouring tea and hammering nails, or, in the context of a pandemic, swabbing noses and diagnosing infections.

The Reality of Children's Play

What do children do in their free time? The answer depends on where they are growing up. Children everywhere play, but the quantity and quality of their play differs by culture. The children who spend the most time playing—and the most time playing pretend games in particular—are children growing up in Western, industrialized societies.[5] Children growing up in other societies play less and work more, particularly children in traditional, small-scale societies.[6] They spend most of their day observing and assisting the work of adults. When left to their own devices, they often emulate that work in their play.

Yucatec Mayan children spend most of their day doing chores, running errands, or taking care of younger siblings. When they play, they usually construct or manipulate objects rather than pretend. If they do engage in pretense, they pretend to do real activities like hunting, cleaning, or selling produce.[7] Aka children living in the

Congo basin exhibit a similar profile of free-time activities. They spend only a quarter of their day playing but rarely engage in pretense, other than simulating the work of adults. Their most frequent form of play is manipulating physical objects and roaming the local environment. Children in the nearby farming community of Ngandu spend more time playing and also more time engaged in pretense. But even Ngandu children's pretense consists mainly of pretending to do work.[8]

In one expansive study of what children do in their free time, researchers surveyed 2,400 mothers from sixteen countries and found that pretend play is surprisingly rare.[9] Only a quarter of the mothers reported that their children engage in imaginative play, and only a tenth reported that their children are happiest when doing so. The activities mothers reported to be most common were watching television and playing outside, which were also associated with the highest levels of happiness. It's possible that children's outside play is imaginative, leading mothers to underestimate the time their children spend pretending. It's also possible that children would spend more time pretending if they were not allowed to watch television. But the fact remains that parents do not observe a lot of pretend play, and the lack of such play has been noted in societies as diverse as France, Brazil, Turkey, Morocco, South Africa, India, Indonesia, and China.

One reason children do not engage in pretend play, despite adults' belief that they do, is that children would rather engage in real activities. When researchers asked preschoolers whether they would rather ride a horse or pretend to ride a horse, they overwhelming chose to ride an actual horse.[10] They also chose to bake cookies, cut vegetables, feed a baby, talk on the telephone, and go fishing over just pretending to do those things. When asked why, they claimed it would be more productive, more informative, and simply more fun. They also asserted that they are capable of accomplishing real tasks. "Cause I can do it myself!" exclaimed one child.

Preschoolers' preference for real activities is remarkably robust. They are twice as likely to endorse real activities over pretend ones when reading about these activities in a book, and they spend twice as much time playing with real tools over pretend versions of those tools when given both options.[11] The toy industry makes billions of dollars each year selling parents pretend versions of real objects, like toy dishes and toy phones, but children are not fooled. They would much rather play with real dishes and real phones, as parents quickly discover when children get their hands on such objects. The pretend versions are fun to play with at first, when fresh out of the box, but children quickly tire of them and search for the real thing instead.[12]

Parents in traditional, small-scale societies seem more attuned to children's preference for real activities over pretend ones. They do not encourage pretend play, nor do they provide children with props for conducting it.[13] Instead, they provide children with real tools and encourage them to accomplish real goals.[14] Matsigenka children in the Amazon basin use machetes to cut wood by age three. Maniq children in Thailand use knives to skin animals by age four. Okinawan children in Japan use sickles to peel sugar cane by age five. Sometimes parents in these societies provide children with miniature versions of adult tools, but these versions are functional and intended to be used for real.[15]

Westerners who observe toddlers in non-Western cultures carry machetes or play with fire are often aghast, but the parents of such toddlers would be aghast by the Western practice of sequestering children from adult work and expecting them to play pretend games instead. These parents view participation in adult work as critical to learning and development. "How can you learn to use a knife if you do not use it?" asks a Dusun father from North Borneo, when queried on why he was comfortable giving his toddler a knife.[16]

In sum, the idea that children instinctively and habitually immerse themselves in pretense is itself a pretense. Children spend much less time pretending than commonly believed, and when they do engage in pretense, it is typically for the purpose of rehearsing real-world activities, particularly the work of adults. That said, there is a difference between how pretense is typically used and how it could be used—as a means of vicarious exploration. Pretense makes it possible to learn about the world without muddying our hands in messy details or tying them with practical considerations. We will explore this possibility in the following sections by considering the role of pretense in learning and reasoning, as well as its role in socializing and collaborating.

Fancy but Not Fanciful

Pretense, at its heart, is a symbolic activity. We posit ideas that symbolize, or represent, something else. A child who pretends to diagnose her teddy bear with coronavirus posits several such representations: her teddy bear represents a patient, she represents a doctor, and her diagnosis represents an established medical activity.

Although pretense may appear simple on its surface, it actually requires some sophisticated mental accounting.[17] The child must catalogue the various symbols at play—"doctor," "patient," "diagnosis"—and incorporate those symbols into a unified script. She must then elaborate the script in ways that honor the relations among the symbols as well as the properties of what they stand for. Her pretense would be incoherent, for instance, if she stipulated that the patient diagnose the doctor or that the diagnosis be made on the basis of a wrestling match. Additional effort is required to maintain a division between the symbols and the objects used to instantiate them, such as a division between the size and sentience of a teddy bear and that of a coronavirus patient. And even more effort is required to communicate and coordinate these activities if the pretend play involves other participants.

The sophistication of this activity is underscored by a comic from poorlydrawnlines.com, featuring a mother and her baby at the doctor's office. The doctor talks to the mother while her baby plays with a toy telephone in the background.

Doctor: I'm so sorry to tell you this, Mrs. Rogers, but your baby is insane.

Mother: What?

Doctor: He appears to believe this telephone is real, but you can't even call anyone with it.

Mother: Oh my God.

Doctor: And the buttons are just silly faces.

Mother: Oh my God!

Pretending requires keeping track of the pretend world while also sequestering the pretend world from the real world. These abilities appear to emerge spontaneously, without guidance or instruction. Children as young as fifteen months are able to establish and accomplish pretend goals, such as feeding a toy dog or washing a doll's hands.[18] Two-year-olds are able to distinguish pretend scenarios from real scenarios and act appropriately in each.[19] If, for instance, an adult holding a teddy bear asks a two-year-old for some blocks, the child will give the blocks to the adult if the request is made in a normal voice but will give the blocks to the bear if made with an exaggerated voice, interpreted as the bear's.

Children's early proclivity for pretense is accompanied by an early appreciation of how to track pretend actions and manipulate pretend objects. Pretend activities could, in principle, unfold in any way, but activities that unfold bizarrely would not be informative to contemplate or easy to coordinate. A child who pretends that applying soap and water to her doll's hands makes them dirtier and applying a towel makes them wetter would not learn anything useful about hand washing, nor would she be able to coordinate these activities with a peer, as the two would be unable to align their expectations about the activities' outcomes.

And studies show that children's expectations do align.[20] Children as young as two expect that pretend water will be transferred from one cup to another when someone pantomimes pouring and that pretend chalk will be transferred from one bowl to another when someone pantomimes scooping. They expect that a doll who touches pretend water will be wet but a doll who touches pretend chalk will be powdery. And they can keep track of which dolls are wet and which dolls are powdery when multiple dolls are involved.

Young children also expect that others' pretend play will follow the logic of real-world transformations. They expect that, when playing tea, guests at the tea party should drink from cups only if tea has been poured into them.[21] And they expect that pretend scenarios involving animals should honor basic facts about those animals, such as the sounds they make.[22] If they watch a pretend scenario in which a toy duck says "oink," they will balk at this event, exclaiming, "That's not what ducks say; ducks say quack." Some three-year-olds simply yell, "Stop!" They protest even when the scenario is framed as a "made-up story" that happened "once upon a time." Ducks are supposed to quack in fairy tales just as in real life.

Children's insistence that pretense should conform to real-world regularities is both a blessing and a curse. It's a blessing in that children can use pretense to practice real-world skills, like pouring liquids and washing hands, and they can easily coordinate their pretense with others. But it's a curse in that they have difficulty generating ideas that do not conform to those regularities.

Children's fixation on real-world regularities is particularly evident from their drawings. When children draw a house, they tend to draw a prototypical house: a box with a triangular roof, rectangular chimney, two windows, and a door. They produce these drawings even when prompted to draw an unusual house—a "house that does not exist." Older children respond to this prompt by drawing

Figure 8.1 Children asked to draw "a person that doesn't exist" typically draw a normal person with an extra feature or a missing feature. (Adapted from Karmiloff-Smith 1990) (Courtesy of the author)

a house with an unusual shape or by placing their windows and doors in an unusual configuration, but preschoolers are unsure of how to respond. They either ignore the prompt or draw a standard house and then erase its windows or its chimney.[23]

It's not just houses that preschoolers find difficult to reimagine. They also falter when asked to draw an animal that doesn't exist or a person that doesn't exist. In one study, researchers asked children of various ages to draw a person that doesn't exist on four occasions over four months.[24] They found that younger children, around age six, drew people that differed only moderately from real people, such as a person with a large head or a person with only one leg. Children who were shown examples of more exotic drawings, such as a person with four arms or a person with a leg protruding from their shoulder, produced more exotic drawings themselves but only in the short term. After a few months, they returned to their original strategy of drawing a real person and then deleting or manipulating one of its features.

This strategy of generating something new by tinkering with something familiar does not change with age. If adults are asked to draw an animal that doesn't exist, most will draw an animal that does exist but then tinker with its features.[25] We might draw a deer, for instance, but add antennae, or draw a shark with spikes or a gerbil with insect eyes.

The similarity between these made-up animals and real animals can be seen at multiple levels, from similar appendages to similar body plans. Consider, for instance, what happens when adults are asked to draw two made-up animals: one with feathers and one with scales. The creature with feathers ends up looking a lot like a bird, and the creature with scales ends up looking a lot like a fish. The birdlike creature usually has wings and feet, and the fishlike creature usually has gills and fins. A creature with feathers doesn't have to have wings; it could have fins. But feathers prime the idea of birds, and birds prime the idea of wings, just as scales prime the idea of fish, and fish prime the idea of fins.

Aside from specific features, the birdlike creature and the fishlike creature will resemble earth animals in their higher-order properties as well, such as symmetry and the location of sensory organs. We may add extra wings to our birdlike creature, but we would add them in pairs, balanced across the creature's midline. And we may add an extra eye to our fishlike creature, but we would add the eye to its head rather than its belly or back.

One could quibble over the significance of such findings, given that participants were asked to draw animals and animals have defining features, like wings and eyes, that set them apart from other objects. Can people be faulted for including real-world features if those features are necessary for identifying what the drawing

Figure 8.2 When adults are asked to draw animals from another planet, they draw animals that largely resemble those found on earth. This pattern is particularly pronounced when asked to draw animals that retain a familiar feature, such as scales (bottom left) or feathers (bottom right). (Adapted from Ward 1994) (Courtesy of the author)

is? This objection highlights a central tension in what it means to pretend—a tension that makes pretense paradoxical. An idea must be decoupled from reality to count as pretend, but it must include some amount of real-world content to count as relevant or meaningful. The line between too much real-world content and too little varies by context, but the findings reviewed here suggest that we are prone to err on the side of too much, especially when we are young.

Enriching but Not Enlightening

One benefit of the close connection between pretense and reality is that pretense can be used as a vehicle for learning. If children watch a puppet show in which a dog runs away from a raccoon, they assume that dogs really are afraid of raccoons.[26] If they watch someone pretend to use a kitchen utensil to flatten dough, they assume the utensil really is for flattening dough.[27] Such learning persists long after the pretend demonstration.[28]

Children thus treat pretend demonstrations as no different from real demonstrations so long as the demonstrations do not conflict with their prior knowledge.[29] Conflicting information is rejected, just as it is when children receive conflicting information in the form of testimony. Children will not accept that a dog could drive a truck, whether they are told about a truck-driving dog or watch a dog drive a truck in a puppet show. Pretense, like testimony, is accepted only when it conforms to pre-existing expectations.

In this way, pretense can aid learning. Might it aid other aspects of development as well? Pretense requires manipulating and coordinating symbolic representations, so it's plausible that pretense might facilitate the development of other symbolic skills, such as language, reasoning, problem-solving, or executive function. Dozens of studies have explored this possibility, but few have corroborated it.[30] Pretend play is sometimes correlated with symbolic skills, but there is no evidence that it facilitates or enhances them. Rather, the evidence appears to be coincidental: situations that foster cognitive development also foster pretend play. When controlled experiments have been conducted, they reveal that children encouraged to engage in pretend play show no cognitive advantages over those engaged in other activities.

Pretend play is, of course, just one form of play. Other forms include free exploration, object manipulation, and competitive games. Might these other forms have lasting influences on cognition? Studies suggest that play in general has the same cost–benefit profile as pretend play in particular. The benefit is that we may uncover information that will enrich our understanding of the situation, but the cost is that we are unlikely to discover information that might challenge our understanding and require us to learn new concepts or principles. Play is enriching but not enlightening, at least not without external guidance.

One way that play is enriching is by revealing hidden affordances. Preschoolers will play with a new toy until they have determined what it does and how it works.[31] If, for instance, preschoolers are given a toy box that plays music when it comes into contact with patterned beads, they will test the beads one by one to determine which patterns produce music and where on the box they need to make contact. If preschoolers are given boxes with secret contents, such as pencils or marbles, they will shake the boxes to determine what is inside and how many there are.[32] The time they spend shaking these boxes is neither constant nor random but calibrated to the informativeness of their shakes. The harder it is to tell what's inside the box, the more preschoolers shake it.

Play reveals new information about the objects we're playing with. But if that information is unexpected or anomalous, we are prone to overlook it. Anomalies revealed in the act of play are no more useful than anomalies revealed through testimony or observation, as discussed in earlier chapters. Science museums are filled with displays intended to teach physical principles through play, but children rarely play with them in ways that might reveal the relevant principles. And if they do, they rarely glean those principles from their observations.[33] Children who roll balls down ramps, for instance, learn that the height of the ramp influences the speed of the ball, as expected, but they rarely learn that the weight of the ball makes no difference.[34] Children who drop blocks in water learn that heavy blocks sink faster than light blocks of the same size, as expected, but rarely learn that the blocks' shape matters, too.[35]

The situation is no better for educational video games, designed to foster learning through play. Gamers can learn to reap a game's rewards, by completing discrete tasks or meeting discrete targets, without learning anything about the situation the game was meant to model. Consider the game of Electric Field Hockey, designed to teach the physics of electrical interactions.[36] The objective of the game is to move charged particles into a goal by placing other charged particles along its path. The game instantiates several electromagnetic principles: that a particle's acceleration is proportional to the force applied by other particles, that this force increases exponentially as the particles approach one another, and that the force exerted by multiple particles is a linear combination of their individual forces.

But students who play Electric Field Hockey rarely learn any of these principles. Instead, they spend their time jerry-rigging complicated trajectories through trial-and-error placement of individual particles. The game designers expected students to learn the relations between force, distance, and acceleration and then use those relations to plan simple paths that move the target particle—the puck—with as few additional particles as possible. Yet most students approach the problem with brute force, shoehorning the puck across the screen by piling one particle on top of another.

Students who play Electric Field Hockey fixate on the game's goals rather than its underlying principles because they have no incentive to heed such principles if they can succeed without them. Similar results have been observed for games designed to teach the principles of projectile motion, pendulum motion, chemical reactions, and thermodynamics.[37] The best way to teach students principles is not creating situations governed by those principles and then encouraging them to play but telling students what the principles are, point-blank.[38]

Play may be fun, but its efficacy as a teaching strategy pales in comparison to that of direct instruction. Studies have found that direct instruction leads to better learning outcomes than self-directed exploration for a wide range of topics, contexts, and students.[39] As one team of educational psychologists concluded, "There is no body of research supporting this technique. In so far as there is any evidence from controlled studies, it uniformly supports direct, strong instructional guidance."[40]

A wrinkle to these findings is that play can be beneficial if it is guided rather than self-directed, as noted in chapter 4. In fact, guided play is often more effective than direct instruction.[41] Guided play combines the intrinsically enjoyable aspects of exploration with the pedagogically valuable aspects of instruction, yielding experiences that are at once engaging and enlightening.[42] The success of guided play is nicely illustrated by the game discussed earlier, Electric Field Hockey. When students play the game in whatever way they choose, they neglect the physical principles animating game play, but their attention can be drawn to those principles by asking the participants to shoot the puck along a predetermined path. Fitting the puck's motion to the path requires attending to the inverse relation between force and distance as well as the additive relation among charges. Sure enough, students who play the guided version of Electric Field Hockey learn the physical principles that self-directed students overlook.[43]

Guided play has proven more effective than self-directed exploration for learning a wide variety of topics, from complex topics like electromagnetism to simple topics like identifying geometric shapes or predicting the motion of interlocking gears.[44] These successes indicate that play has a legitimate place in the classroom, but it's a place that must be prepared and curated. When we play, we tend to explore the conceptual terrain we already know well. It takes guidance to push us into new terrain, capable of illuminating new concepts.

Following the Rules

As a symbolic activity, pretense has a natural affinity to rules. Symbols are only useful, after all, if they are manipulated in ways that conform to what they symbolize. This conformity explains why children who play hospital assign doctors the task of making diagnoses and assign patients the task of being diagnosed. It also explains why children are able to share pretense so easily. Children who play hospital together do not have to work out the rules beforehand. They know that the rules governing real hospitals will govern their pretend hospital as well.

Pretense is about following rules, not breaking them. In fact, when rules are broken, children get upset. They protest the infraction and try to correct it. Children as young as two can easily pretend that a wooden clothespin is a knife and a plastic clothespin is a carrot and then use the wooden clothespin to chop the plastic one. But if someone joins the game and starts using the plastic clothespin to chop the wooden one, they exclaim, "No! Not like that!" and insist "*this* one is a carrot and *that* one is a knife."[45] Children's protests reveal not only that they believe rules must be followed but also that they understand rules must be set. They will chastise a partner for using the pretend knife as a clothespin (which, in fact, it is), but they will not chastise her if she is an outsider who has yet to join the game and does not know its rules.[46]

Children's affinity for rules is even more apparent in pretend games where the rules are completely arbitrary. Consider the game of daxing. It's played with a block, a board, and a rake. The objective is to move the block from one side of the board into a gutter on the other side. People who dax use the rake to push the block into the gutter. Of course, there are other ways to get the block into the gutter. You could bat it in, or you could lift the board and slide it in. People who dax do not appear to move the block in these other ways, but are they okay?

Figure 8.3 Young children taught that "daxing" is using a rake to push a block across a board are appalled to observe someone move the block across the board in another way. (Courtesy of the author)

Two-year-olds are adamant that they are *not*.[47] They chastise people who attempt these other ways to move the block and show them the correct way to dax, even when they themselves were introduced to daxing only moments prior. Two-year-olds do not need to be told that daxing has one unbreakable rule; they infer this rule from a single demonstration, similar to how children infer the "true" function of a tool from a single demonstration. (Daxing, by the way, has three rules. The third is that you push the block with a rake; the first two are that you do not talk about daxing.)

Children are not only upset when the rules of a game are broken but also hold the rule breaker in disdain. When preschoolers catch someone breaking a rule, they report that they don't want to play with that person again, either in the same game or a different game. They also report that they don't want to be friends with him or share candy with him, and they think that he's mostly mean. Preschoolers develop these negative attitudes even when the rule breaker ends up helping the child win the game.[48]

The rule breakers that children hold most in disdain are those who should know better: people who are part of their social group and should therefore be familiar with the rules. If three-year-olds play a game of daxing with someone who breaks the rules, they will chastise the rule breaker more vehemently if she lives in the same community and speaks with the same accent than if she lives in a foreign land and speaks with a foreign accent.[49]

Children of this age appear to understand that the rules of a game are established by consensus and that people who did not participate in this consensus—either directly, in creating the game, or indirectly, in inheriting the game from their community—can be cut some slack if they happen to break the rules. Children will also cut rule breakers some slack if they perceive that consensus about the rules is not absolute—if, for instance, game players openly argue about the rules.[50] Games are a form of shared intentionality, where establishing and enforcing rules is as important as following them. Half the fun of daxing is following the rules of daxing. In fact, *all* the fun of daxing is following the rules because daxing is just pushing a block into a gutter.

You may scoff at the simplicity of daxing, viewing it as a game that only three-year-olds could enjoy, but is it all that different from the board games adults play? Aren't games like parcheesi, backgammon, checkers, and chess just elaborate ways of pushing objects across a board? Nothing tangible is accomplished by moving game tokens from one side of a board to the other, and nothing is learned, other than how to move these tokens more effectively in the next game.

Still, adults around the world spend their free time moving tokens around a board, and they have been doing so since the dawn of civilization.[51] Ancient Aztecs played a game like parcheesi. Ancient Indians played a game like chess. Ancient Babylonians played a game like checkers. And ancient Middle Easterners played a game like mancala. Moving objects around boards has widespread appeal because we revel in the act of rule following. Games are an opportunity to collaborate with like-minded rule followers and, in so doing, practice our commitment to shared intentionality.

The philosopher Ludwig Wittgenstein famously argued that there is no single definition of "game," as no definition could capture the varied activities involved in board games, card games, ball games, strategy games, and pretend games.[52] Following Wittgenstein, modern philosophers have analyzed games in many ways: as art, as fiction, as experience, and as performance.[53] These analyses capture different aspects of why people play games and what they gain in the process, but the act of playing a game generally involves rules. Rules establish collective intentionality in a particular space and at a particular time. They create what some philosophers call "a magic circle," where the rules of the real world are set aside in favor of rules that define a new social reality.[54]

Rules are thus integral to pretense, from children's spontaneous episodes of make-believe to the highly organized games played by adults. We care about the rules as much as we care about the actions they define. Without rules, we'd just be pushing objects around boards, sorting cards into stacks, and putting balls into holes, without reason or reward.

Following the Logic

Pretense is commonly believed to be about breaking rules, not following them, and about whimsy, not logic. But neither belief is correct; pretense is driven by rules and logic. We've discussed how pretense conforms to the causal logic of real-world events, but there's another way pretense relates to logic: it can facilitate, and improve, logical reasoning.

Consider the following logic problem: If all milk is black and there is milk on the table, is the milk on the table black? You might be thinking, "Yes, of course. You told me milk is black, so the milk on the table must be black as well." But not everyone would agree. Young children deny that the milk on the table is black because all the milk they've seen is white.[55] Adults who have not attended school say the same.[56] To draw the inference that the milk on the table is black, one must embrace a premise known to be false, and people who have not had training in this kind of reasoning default to what they know rather than what they have been asked to presume. In other words, drawing logical inferences requires attending to the structure of an argument, not its content, and doing so takes practice, especially when the content conflicts with prior knowledge.

But there's a shortcut to learning the rules of logic: pretense. Young children have no trouble affirming that the milk on the table is black if they are told to imagine that the milk is part of a fairy tale, where funny things happen, or from another planet, where everything is different.[57] Children are twice as accurate at drawing logical inferences from counterfactual premises when they are told to imagine that the premises are true of a pretend context. They will accept, for instance, that Rex the cat can bark if Rex is from a planet where cats bark or that Tot the fish lives in trees if Tot is from a fairyland where fish live in trees.

Children are surprisingly willing to draw inferences from counterfactual premises if those premises are embedded in pretense.

They will do so if asked to visualize those premises in their imagination, as well as if the premises are described to them with an exaggerated, make-believe intonation.[58] A week later, they will continue to draw counterfactual inferences, without additional prompting.[59] These interventions work even with two-year-old children, who are just beginning to engage in verbal reasoning, as well as unschooled adults, who have spent a lifetime drawing inferences from facts rather than counterfactuals.[60]

Pretense facilitates logical reasoning in another way as well, by helping us consider alternative possibilities. The type of inference discussed thus far is known as *modus ponens.* It entails extending a property of a category to a new instance of the category, and it typically poses no problems. Problems arise only when the premise of a modus ponens argument contradicts our prior beliefs, as in "all cats bark" or "all fish live in trees," leading us to reject a valid inference.[61] The bias to focus on believability over logic is known as belief bias, and this bias works in the opposite direction as well. We are prone to accept logically *invalid* inferences if they accord with prior beliefs.

Consider the following argument: "If you go swimming, you will get wet; you are wet; therefore, you must have gone swimming." This argument strikes many people as valid, but it's not. It's an example of "affirming the consequent," where we erroneously assume that the antecedent of a conditional statement (going swimming) must be true if its consequent (getting wet) is true. But stop for a moment and consider all the ways you could get wet without going swimming. You could have taken a shower, walked through a sprinkler, got caught in a downpour, or fallen in a lake. Although swimming makes a person wet, it's not the only way a person can get wet, and thinking of alternative possibilities allows us to see the fallacy of that inference.

The reason we affirm the consequent when we shouldn't is that we do not regularly generate alternative possibilities to check the validity of our inferences, but we can be encouraged to do so with pretense. When we are told "if you go swimming, you will get wet," the possibility of getting wet from swimming is emphasized over other possibilities. But if we are good at brainstorming alternative possibilities, we are less swayed by the possibility suggested in the argument's antecedent. This relationship, between brainstorming and logical inference, has been demonstrated in both children and adults; those who are good at brainstorming are less likely to affirm the consequent when drawing conditional inferences.[62]

For example, most five-year-olds draw the wrong inference when answering this question: "All dogs have legs; a friend of mine has an animal with legs; is it certain the animal is a dog?" They say it is, failing to recognize that cats, rabbits, and gerbils also have legs but are not dogs. But five-year-olds' accuracy is dramatically improved if they first take part in a brainstorming session, where they brainstorm different gifts to give a friend or different ways to make noise.[63] Putting children in the mindset of generating alternative possibilities cues them to generate alternatives to the antecedent (being a dog) that still satisfy the consequent (having legs). Critically, this intervention works best if children generate possibilities on their own. Children who are given help during the brainstorming session still end up affirming the antecedent.

Pretense thus sharpens our logical reasoning skills. In fact, the activity of reasoning logically—of drawing out the implications of counterfactual claims or generating alternatives to a suggested conclusion—may be what makes pretense fun. Logical inference sounds tedious, but it's actually what gives pretense its substance and coherence. It's what elevates pretense from a random activity to a meaningful exercise.

Psychologists have noted several ways in which pretense does not just facilitate logical reasoning but is constitutive of it. Consider the reasoning behind what-if questions and how-might questions. To answer what-if questions like "What if milk were black?" or "What if cats could bark?" we have to decouple a hypothetical possibility from reality, explore what inferences follow from the hypothetical, and then quarantine those inferences from our real-world beliefs.[64] To answer how-might questions, like "How might I make milk black?" or "How might I teach my cat to bark?" we have to decouple our hypothetical goals from our current goals, explore different solutions to those goals, and then imagine how to apply those solutions to real-world contexts.[65]

These activities are fun. In fact, they are often more fun than solving real problems, with real costs and real constraints. Pretense is a world of possibilities we can explore whenever we want and however we want. But the outcome of the exploration is determined by logic, not desire.

In sum, pretense has qualities the average person would associate with deliberation rather than imagination. It's realistic, not fantastical; structured, not capricious; and rule governed, not rule breaking. The contrast between the way pretense is and the way it's perceived is nicely captured in an exchange between two characters on the sitcom *Brooklyn 99*—Sergeant Jeffords and Captain Holt—who argue over the best way to build a model railroad.

> *Jeffords:* Check out Jeffords Junction. It's got an ice cream shop, a '50s diner, and a Mexican restaurant with a sombrero on it.
>
> *Holt:* What happens to the sombrero when it rains? I see no drains. Does it just fill up with water until the building collapses, killing everyone inside?
>
> *Jeffords:* I don't know; I'm just trying to build a fun world. That's what model trains are all about: imagination.
>
> *Holt:* No, they're about accuracy. People like to see the world around them exactly as it is, but smaller.

The straightlaced Captain Holt is intended to be the foil, expressing a view of pretense that seems patently wrong. But Captain Holt is right: the pretend worlds we create and play within are usually close replicas of the real world. The closer the replica, the more reliably we can learn from our pretense and the more easily we can share it with others. The machinery of pretense can be co-opted to contemplate fantasy worlds and illogical ideas, but its true power lies in uncovering the hidden logic of reality.

9 Fiction

Expanding Our Social Imagination

If you're like most people, you're a fan of Harry Potter. But guess what: the Harry Potter books are full of lies. There's no such place as Hogwarts, no such person as Dumbledore, no such animals as hippogriffs, and no such sport as quidditch. Magic doesn't exist. Magical creatures don't exist. Even Harry Potter doesn't exist. And yet we willingly spend leisure time engaged with these lies—a lot of time.[1] The Harry Potter books have sold over 500 million copies, and each averages around 500 pages. If each page takes about a minute to read and the books have been read about 500 million times, humanity has spent 250 billion minutes reading Harry Potter—or over 475 thousand years. And that figure doesn't include the time spent watching Harry Potter movies, exploring Harry Potter websites, or visiting Harry Potter theme parks.

All works of fiction—novels, plays, sitcoms, movies, musicals, operas—are lies. We know they didn't happen and often could never happen, but we still find them worthy of our time, often more worthy than other, more practical activities. Why?

The answers are as varied as fiction itself. Fiction can educate us on the risks and benefits of actions we have not (or cannot) undertake, as well as their moral or logical consequences.[2] It can appease ever-present drives, such as the drive to form coalitions, the drive to attain status, and the drive to find mates.[3] It can provide an escape from current pains and a window onto future pleasures.[4] And it can sharpen our interpersonal skills by simulating social interactions and evoking social emotions.[5] "What do stories do?" asks literary critic Jens Eder. "They make children go to sleep and soldiers go to war."[6]

Underlying these varied functions is a common form: a model of the social world built to be experienced and explored. Like pretense, fiction is an exercise in decoupling facts from counterfactuals, to reason about the latter in order to learn more about the former. Also like pretense, it has the paradoxical quality of being separate from reality while also closely tied to it. We could potentially experience anything in fiction—what it's like to live inside a deep-sea vent, what it's like to age in reverse, what it's like to be a cockroach—but, as research reveals, we prefer fiction that hews close to what we already know and have already experienced. Even when we contemplate genuinely fantastical worlds like Harry Potter's, those worlds are far more similar to the real world than they are different.

In this chapter we will explore what makes fictional worlds appealing, what we expect those worlds to be like, and what we learn from those worlds—what ideas we export from fiction into reality. In all cases, fiction hews closely to fact. We prefer realistic scenarios to unrealistic ones, expect the events in those scenarios to unfold similarly to real events, and learn from fictional scenarios only if they map easily onto reality. Imagination is a prerequisite for engaging with fictional worlds, but that engagement is constrained by our less-than-imaginative beliefs about everyday life.

And consistent with the research reviewed in previous chapters, the people most constrained by reality when engaged with fiction are children. Many beloved works of fiction were written expressly for children—the Harry Potter books, the Narnia books, the Oz books, the Wonderland books—but children don't particularly like fiction, at least not initially, and they rarely learn from fiction, at least not the lessons the authors intend them to learn. Children's fiction would be much less imaginative if children wrote it themselves.

An Acquired Taste

If I were to read you a story, would you rather hear a story about a boy who has a lot of brothers and sisters or a story about a boy who lives on an invisible farm? Most adults pick the second story, but most preschoolers do not. They are as likely to pick the story about a large family as the story about an invisible farm. Although children's books often contain fantastical elements, like invisible farms, studies find that fantasy appeals more to older audiences than younger ones.[7]

In fact, younger audiences are averse to fiction in general. Given the option between a "true story" about a boy who finds a dinosaur bone and a "make-believe story" about a boy who finds buried treasure, preschoolers reliably pick the story about finding a dinosaur bone. If the options are reversed—the story about buried treasure is described as true and the story about a dinosaur bone is described as make-believe—preschoolers now pick the story about buried treasure.[8] Just as preschoolers prefer realistic activities to make-believe activities when playing, as noted in the previous chapter, they prefer realistic stories to make-believe stories when reading.

Children's preference for realistic stories can also be seen in the events they select when constructing their own stories.[9] Researchers have explored these preferences using a choose-your-own-adventure paradigm, where children are introduced to a character and a setting and asked to complete the story by selecting between alternative story events, one ordinary and one fantastical. Preschoolers in one study were introduced to Moe, a boy in elementary school, and asked to complete Moe's story by deciding whether Moe should have one brother or sixty-two, whether he should drive to school or fly, whether he should play catch with his friends or play a telekinetic ball game, and whether he should go to sleep at the end of the day or stay awake all night. Preschoolers reliably selected the ordinary events. They selected the ordinary events even when Moe was replaced with Zoltron, an alien from outer space. Zoltron's status as an imaginary creature did not cue them, or encourage them, to select the fantastical events instead.

Preschoolers' preference for realism is so strong that they will override their knowledge of fictional genres to select realistic events over fantastical ones. Preschoolers know, for instance, that stories that take place in outer space include aliens rather than dragons, and stories that take place in a castle include dragons rather than aliens. And they can use that knowledge to select genre-appropriate events for such stories, selecting a spaceship over a carriage as the appropriate vehicle for transporting children who live in space but making the opposite selection for children who live in a castle. However, if given the option of a bus, preschoolers select the bus over either a spaceship or a carriage. Although they prefer genre-appropriate events to genre-inappropriate ones, they prefer realistic events to both.[10]

Children's fixation on ordinary events extends even to realistic fiction. When preschoolers are asked to decide between two realistic events, one typical and one atypical, they opt for the typical event. And they do so even when earlier events in the story were atypical. Preschoolers who hear a story about animals eating unusual foods, such as a monkey eating an apple and a squirrel eating a pear, will end the story with a dog eating a bone rather than a dog eating peas.[11]

Thus, contrary to popular belief, young children do not love fantasy stories. They prefer realism to fantasy, nonfiction to fiction, and typical stories to atypical ones. They easily engage with fictional stories—tracking their characters, following their plots, even learning their conventions—but they prefer stories that confirm, rather than challenge, their knowledge of real-world events. If *Clifford the Big Red Dog* were a choose-your-own-adventure story, children would chose that Clifford was medium-sized and brown.

Based on Real Events

Fiction is an acquired taste, but even those who have acquired it never truly abandon their fondness for realism. Consider J.R.R. Tolkien's experience interacting with fans of *The Hobbit* and *The Lord of the Rings,* set in the fictional world of Middle Earth. Middle Earth is populated with magical beasts, like dragons and balrogs; magical people, like elves and dwarves; and magical artifacts, like enchanted rings and sentient swords. These are the elements of Middle Earth that would seem to make it uniquely attractive, drawing readers in and keeping them engaged. But it was the mundane elements of Middle Earth—its lands, languages, and cultures—that fans were most eager to discuss with Tolkien.

"Most people want more (and better) maps," Tolkien reported to a friend. "Some wish more for geological indications than place-names; many want more specimens of Elvish, with structural and grammatical sketches. . . . Musicians want tunes and musical notations. Archaeologists enquire about ceramics, metallurgy, tools and architecture. Botanists desire more accurate descriptions of the [plants]. Historians require more details about the social and political structure of Gondor, and the contemporary monetary system."[12]

It's not just fans of Middle Earth who are curious about the mundane details of fictional worlds. Fans of Star Wars obsess over the names and locations of planets in the Star Wars galaxy. Fans of Harry Potter obsess over the rules and strategies of quidditch.

And fans of Pokémon obsess over the evolution and taxonomy of Pokémon species.[13] We know that fictional worlds do not exist, and often could not exist, yet we still assume they are grounded on a bedrock of fact.

Why do we make this assumption? The answer is the same as to why pretend play is so realistic: we use our expectations about the real world to structure our engagement with fictional worlds. We are willing to suspend some expectations but not many. And our willingness to suspend an expectation depends on how deeply it is entrenched in our knowledge of real-world events.

Consider a story about a man who wears a cape, is resistant to pain, and can fly. He goes to a park and saves a child's kite from flying away. It's not the most exciting story or the most detailed, but it still elicits expectations about the world in which it takes place. Chances are you think that math works the same in this fictional world as it does in the real world. You probably also think that science works the same, but you might be less certain about the world's history or geography. It would surprise you to hear that the people in this fictional world do not have hearts, but it wouldn't be so surprising to hear that the capital of the United States is Chicago.[14]

We have an intuitive grasp of the flexibility of real-world regularities, and we use that flexibility to fill in the details of fictional worlds that have been left unspecified. The less flexible the regularity, the more we expect it to hold, even when other regularities are explicitly violated. A flying man nullifies our confidence that US geography is still organized in the same way but not our confidence that two plus two still equals four.

An interesting middle ground is our expectations about social life. While we generally think that picking your nose would be rude in a fictional world, we are less confident of that expectation than the expectation that people still have hearts or that two plus two still equals four. And we have good reason: authors frequently use fiction to expand our sense of social possibility, as we shall discuss later. But for every aspect of social life that an author might change, there are dozens more that remain unchanged.

George R. R. Martin's book *A Game of Thrones,* for instance, challenges readers to imagine a unique social order organized around flying dragons and frozen zombies, but the structure of this society conforms to familiar social relations. There are dozens of characters in this fictional world, yet each maintains a social network typical of real social networks in size and structure. In fact, the connectivity of these networks reflects the characters' social status, as does the connectivity of real-world social networks.[15] Similar patterns have been observed for the characters in Shakespearian dramas.[16] Othello may challenge our expectations about race and class, but his social interactions conform to our expectations about cliques and alliances.

Perhaps the most salient example of how we import the real world into fictional worlds comes from fiction involving magic. Magic, by definition, violates real-world causality, yet we still apply causal expectations to magical events. Take the magical transformations in Ovid's *Metamorphoses,* where sailors are turned into pigs and nymphs are turned into trees, or those in *Grimms' Fairy Tales,* where donkeys are turned into children and straw is turned into gold. These transformations follow a causal logic. Animate entities tend to turn into other animate entities, but inanimate entities tend to remain inanimate.[17] If the boundary between animacy and inanimacy is crossed, the distance between the relevant categories is still minimized. When humans are transformed, they are most likely to turn into animals, next most likely to turn into plants, and least likely to turn into inanimate objects. Rarely are humans ever turned into liquids or immaterial entities like shadows or rainbows.

Magic spells also follow a causal logic. Magic spells violate physical laws, and the more foundational the law, the more difficult we view the spell. Conjuring a frog out of thin air strikes us as very difficult because doing so would violate a foundational law of matter—that objects can neither be created nor destroyed. Doubling a frog's size, on the other hand, strikes us as less difficult because frogs cannot double in size instantaneously but they do double in size over the course of normal growth.[18]

Our intuitions are ordered by when, developmentally, we came to understand the physical law violated by a magic spell. Spells that violate physical laws learned early in development are viewed as more difficult than those learned later.[19] The law that objects cohere as they move is understood early in life, as early as three months of age, whereas the law that objects fall if unsupported is not understood until later, typically around seven months. Accordingly, a spell for dividing one frog into two strikes us as more difficult than a spell for levitating a frog off the ground.[20]

Magic spells also follow another kind of logic. They violate some causal expectations but conform to many others. Levitation spells, for instance, violate the expectation that unsupported objects fall but conform to the expectation that the amount of support an object requires depends on its weight. When levitation is part of a fictional world, the characters typically learn to levitate light objects before moving on to heavier ones, as when Harry Potter learns to levitate feathers before books or when Luke Skywalker learns to levitate rocks before starships.

My colleagues and I have investigated this intuition by asking people to rate the difficulty of different spells that violate the same core principle.[21] The cover story was that Harry Potter's school, the Hogwarts School of Witchcraft and Wizardry, is introducing new spells into its curriculum, and the teachers need guidance on how to sequence them. The participants were shown pairs of spells and asked to decide which is introductory and which is advanced. One such pair was a spell for making a basketball float in the air and a spell for making a bowling ball float in the air. Logically, it shouldn't matter how heavy the ball is if physical contact is no longer required to keep it aloft, but our participants judged that a bowling ball would be more difficult to levitate. We suspend our expectations about support but continue to apply our expectations about weight.

This pattern holds for different kinds of causal violations, from physical to biological to psychological. We think it would be harder to walk through a brick wall than through a wooden wall, harder to grow an extra eye than an extra toe, and harder to teach a cow how to tap dance than how to skip. We honor expectations about real-world events that should no longer be relevant in light of the primary violation at hand. If you can lift an object without touching it, why should it matter whether the object is heavy or light?

It's possible that the participants in our studies judged levitating a bowling ball more difficult than levitating a basketball because they were forced to make a choice. But we've found that giving participants the option of judging the spells equally difficult does not change their inclination to judge the bowling ball spell as more difficult. Likewise, if participants are asked to rate each spell on its own, from not difficult to very difficult, they tend to rate levitating a bowling ball as more difficult than levitating a basketball, even though they could have rated both spells as equally difficult. And if participants are asked to come up with their own spells, they still take extraneous considerations into account. They suggest, for instance, that introductory courses on levitation should focus on objects we frequently pick up, like pencils or spoons; intermediate courses should focus on objects that are more difficult to pick up, like tables or chairs; and advanced courses should focus on objects we can't pick up at all, like cars or refrigerators.

These intuitions are consistent not only across topics and tasks but also across ages and cultures. Preschoolers honor seemingly irrelevant considerations in their judgments of spell difficulty just like adults; they judge that it would be harder to levitate a bowling ball than to levitate a basketball, even though they've had less exposure to depictions of magic in fiction and even less exposure to formal science, where they might learn about the laws of nature violated by magic spells.[22] Adults in China also show the same pattern of judgments, despite exposure to a different literary tradition and a different style of science education.[23]

In a world of magic, anything is possible in principle, but we entertain only some of those possibilities in practice. No one appreciated this quirk of psychology better than Walt Disney.[24] Disney began his career as an animator exploring the possibilities of a fictional world unconstrained by real-world causality. He created surreal cartoons in which clothes jumped from clotheslines and ran around the yard, sausages jumped from grills and danced in a kick line, and pianos turned insolent and bit their players. These cartoons were entertaining but did not have mass appeal; their storylines unfolded in ways that audiences could neither predict nor follow.

To rectify this problem, Disney decided his cartoons had to be "plausibly impossible": they could violate some of the audience's expectations but not too many. His first feature-length movie, *Snow White and the Seven Dwarfs* (1937), embodies this ideal. Snow White talks to forest animals in the movie, but they do not talk back, nor do they levitate or double in size. Snow White's nemesis, the Evil Queen, uses a magic mirror to spy on Snow White and a magic apple to enchant her, but she does not have the power to teleport Snow White, turn her into a frog, or make her implode.

Plausible impossibility may sound logically absurd, but it's psychologically potent. It explains not only why magical worlds have a causal logic but also why some magical worlds are more successful than others, engaging larger audiences for longer periods of time. Snow White's world, for instance, has engaged audiences around the globe for several generations. "Snow White" was first published by the Grimm Brothers in their 1812 collection, and it has since spread across cultures, languages, and media. Other

Figure 9.1 Early cartoons featured surreal events, like hot dogs dancing on a grill and animals removing the top of their head as if it were a hat. Such events violated too many expectations and were eventually replaced with more plausible impossibilities. (Courtesy of the author)

Grimm Brothers tales have proven similarly successful, including "Cinderella," "Rapunzel," and "Little Red Riding Hood."

But the Grimm Brothers collection also included many flops. Have you heard of "The Girl without Hands"? It's a charming tale of a miller who cuts off his daughter's hands in a deal with the devil but later seeks the help of an angel to regrow them. The story involves an enchanted garden, an enchanted letter, and some silver prosthetics—too many implausible events for readers to keep track of. Accordingly, "The Girl without Hands" has fallen into obscurity, along with other lesser-known Grimms' tales such as "The Donkey Lettuce," "The Golden Children," and "Hans My Hedgehog." Indeed, a systematic analysis of the

Grimms' forty-two tales found that the more implausible the tale, the less it has been retold or reproduced since the time of its publication.[25] Magical events may capture our attention but only if tempered with enough real-world regularities to make them plausible.

Gleaning Facts from Fiction

Our expectation that fictional worlds conform, in large part, to the real world allows not only for engagement but also for learning. If we assume that the ideas presented in fictional worlds are true unless otherwise specified, we can incorporate those ideas into our database of real-world facts and concepts.

But is this assumption safe? Isn't fiction false by definition? The philosopher David Lewis considers this question in a famous essay on "Truth in Fiction."[26] He notes that a statement like "Sherlock Holmes lives on Baker Street" is literally false because Sherlock does not exist, but we understand this statement as referring to Sherlock's world, not ours, and thus deem it true. What about a statement like "Sherlock Holmes has two nostrils"? This statement also strikes us as true, even though the author of the Sherlock stories, Sir Arthur Conan Doyle, never specifies how many nostrils Sherlock has. If what defines truth in fiction is explicit stipulation, then we should be ambivalent about the number of Sherlock's nostrils. But we are not ambivalent; we assume that Sherlock has two nostrils just like any other human.

Lewis justifies this assumption in terms of possible worlds. He notes that, when engaged with fiction, "we depart from actuality as far as we must to reach a possible world where the counterfactual supposition comes true (and that might be quite far if the supposition is a fantastic one). But we do not make gratuitous changes. We hold fixed the features of actuality that do not have to be changed as part of the least disruptive way of making the supposition true. We can safely reason from the part of our factual background that is thus held fixed."

Lewis's "possible worlds" view has been supported by empirical studies. When people are asked to gauge the plausibility of counterfactual events, they base their judgments on how far they have to depart from the real world for that event to occur.[27] Cows do not lay eggs in real life, nor do they float in the air, but we tend to think the former is more plausible than the latter because the distance between our world and a world in which cows lay eggs is shorter than the distance between our world and a world in which cows levitate.

Assessment of the similarity between our world and a fictional world underlies our proclivity to learn from fiction. Similarity serves as a proxy for whether the information conveyed in fiction can be treated as truthful. Adults use similarity as a proxy for truth, but so do children. In one study, preschoolers were told about an unfamiliar flower—a popple—that gives people hiccups.[28] They learned about popples in the context of a story, where a character smells one and then starts hiccupping. For some children, the story included only realistic events; the protagonist not only smelled a popple but also drove a car, found a ladybug, climbed a tree, and swam in a pond. For other children, the story included fantastical events, where the protagonist flew through the sky, found a fairy, talked to a tree, and swam through chocolate. When all children were later told that popples are real and that the experimenter smelled a popple on the way to work, only those who heard the realistic story thought the popple may have given her hiccups.

This finding illustrates a general trend in how children learn from fiction.[29] They can learn a wide variety of information—words, facts, concepts, habits, values—and they can learn this information from fiction as well as they can from nonfiction but only if the fiction is realistic. Fantasy elements disrupt learning, and rightly so. How could children trust that popples cause hiccups if popples exist in a world that also includes fairies and talking trees? The distance between that world and our own is just too far.

Children's reluctance to learn new information from fantasy stories extends even to practical information, like problem-solving strategies.[30] The value of a strategy lies in its utility, not its truthfulness, so fantasy elements shouldn't matter if they are irrelevant to the strategy. But they do. Strategies learned from fantasy are rarely transferred to the real world. For instance, when preschoolers hear a story about a monster who moves a pile of marbles by wrapping them in a blanket, they rarely use this strategy to move a real pile of marbles. They remember the monster's strategy but do not use it when faced with a similar problem. Instead, they move the marbles less efficiently, with their hands or with a spoon. What helps preschoolers use the more efficient strategy is replacing the monster with a human; problem-solving strategies modeled in realistic stories are twice as likely to be transferred to the real world than those modeled in fantasy stories.[31]

Fantasy thus serves as a cue for quarantining information, whereas realism encourages us to export that information to the real world. These biases are generally useful, but each has its pitfalls. Just as we are prone to overlook true information presented in a fantasy story, we are prone to embrace false information presented in a realistic one. The realism creates a sense of complacency, where falsehoods slip past our attention in a stream of otherwise truthful information.

If you read a realistic story about Russia, for instance, you might easily overlook a detail implying that Saint Petersburg is the capital of Russia, not Moscow. Your prior knowledge of Russian geography would not necessarily keep you from absorbing this new piece of information if it is embedded in a larger narrative that you accept as true. Later, when asked what the capital of Russia is, you may now recall that it's Saint Petersburg.

This particular falsehood has been implanted in the minds of many research participants exposed to false information in the context of a story and then quizzed on that information later.[32] When people read a story that implies that Saint Petersburg is the capital of Russia, a quarter later repeat that error on a trivia test. Our tendency to accept such errors is not assuaged by encouraging people to read the story more slowly, nor is it assuaged with warnings that the story might contain misinformation. The instruction to "keep in mind that sometimes authors take liberties with facts or ideas so some of the information you will read may be incorrect" has no effect.

It's possible that some of the participants who accepted that Saint Petersburg is the capital of Russia did not know the real capital of Russia beforehand, but researchers have verified that even those who know the capital is Moscow are still susceptible.[33] Around 20 percent of people who answer "Moscow" on a trivia test will change their answer to "Saint Petersburg" following a misleading story. They'll change their answer even if they claimed to be 100 percent confident in their initial answer. And they'll change their answer for all sorts of questions, not just questions about Russian geography. Misleading stories have led people to accept that Benjamin Franklin invented the lightbulb, that the Minotaur was a one-eyed giant in Greek mythology, and that a date is a dried plum.

Acceptance of misinformation has been documented among readers of all ages, but it appears to be strongest in adults, not children. One in four adults will accept a false statement encountered in a story as true, but only one in eight elementary schoolers will do so, and only one in twenty preschoolers.[34] Age differences in susceptibility may be due, in part, to attentiveness; you have to attend to misinformation to accept it. But they may also be due to the expectation that stories are truthful. Although no one expects that fantasy stories are truthful, adults may expect that realistic stories are, at least to a greater degree than children. Why, after all, would an author change the capital of Russia from Moscow to Saint Petersburg for no reason? The more fiction we encounter, the more we expect authors to maintain real-world facts and regularities when crafting fictional worlds, changing them only when necessary.

Reading Minds

Fiction's role in fact-based learning is variable, but there's another way we learn from fiction that is decidedly more robust: learning about others' minds. Fiction allows us to contemplate how people might behave in unusual situations by allowing us to simulate their thoughts, emotions, and intentions. Even fantastical contexts provide fodder for contemplating others' minds in ways that are still useful for understanding real minds.

A wonderful example comes from Frank Herbert's science-fiction classic *Dune*. On the planet where the story takes place, water is scarce. People spend every day extracting water from the environment and conserving it with great care. They use special instruments to harvest moisture from the air; they wear special suits to recapture the moisture in their breath and sweat; and they "render" the dead into pulp, to extract water from their corpse. Against this backdrop of extreme conservation, Herbert describes a scene, tragically omitted from the 2021 movie, in which a new duke hosts a dinner party where he intentionally wastes water to test the loyalty of his court:

> The Duke allowed his voice to trail off on the last line, took a deep drink from his water flagon, slammed it back onto the table. Water slopped over the brim onto the linen. The others drank in embarrassed silence. Again, the Duke lifted his water flagon, and this time emptied its remaining half

> onto the floor, knowing that the others around the table must do the same.
>
> Jessica was first to follow his example. There was a frozen moment before the others began emptying their flagons. . . . She found herself fascinated by what her guests' actions revealed, especially among the women. This was clean, potable water, not something already cast away in a sopping towel. Reluctance to just discard it exposed itself in trembling hands, delayed reactions, nervous laughter . . . and violent obedience to the necessity. One woman dropped her flagon, looked the other way as her male companion recovered it.[35]

As you read this passage, you were probably not dehydrated or even thirsty. But you could still feel what it must be like to be desperate for water—to plan your life around water conservation—and then watch an authority flout those practices, expecting you to follow his lead. What must have gone through the Duke's mind as he dropped his flagon? What must have gone through his guests' minds? How might the Duke have interpreted their reactions? How might the guests have interpreted the Duke's reaction to their reaction?

This calculus is typically what draws us to fiction. We enjoy thinking about what others might be thinking, and fiction provides us with an outlet for this activity. The literary scholar Lisa Zunshine explains fiction's appeal by arguing that our ability to infer others' mental states from their behavior—our theory of mind—is a "hungry" adaptation, in constant need of input.[36] She describes theory of mind as "promiscuous, voracious, and proactive" and notes that it can be stimulated "either by direct interactions with other people or by imaginary approximations of such interactions, which include countless forms of representational art and narrative."

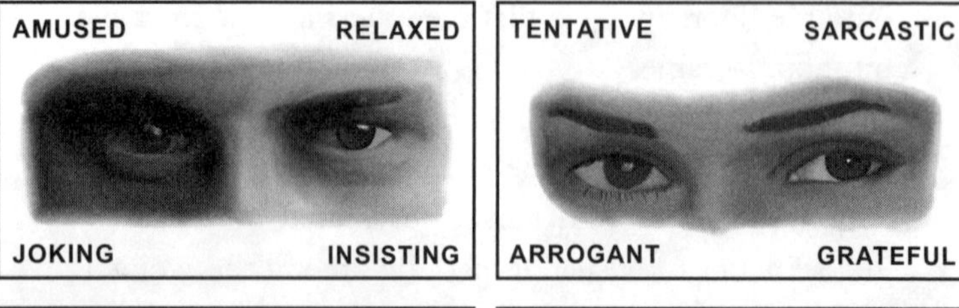

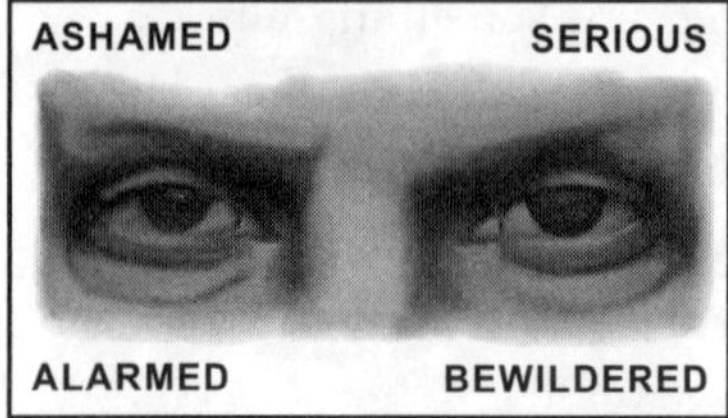

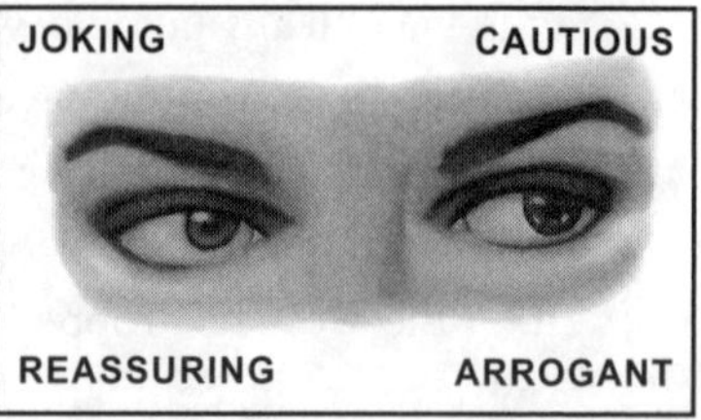

Figure 9.2 People that read a lot of fiction are also skilled at reading minds, discerning, for instance, that the emotions behind these eyes are insisting (top left), tentative (top right), serious (bottom left), and cautious (bottom right). (Adapted from Kynast et al., 2020) (Courtesy of the author)

Fiction not only satiates theory of mind but also instructs it. People who read more fiction exhibit enhanced social reasoning, from identifying others' emotions, to adopting their perspective, to empathizing with their situation.[37] Some of the most compelling evidence for this relationship comes from a test dubbed "Reading the Mind in the Eyes."[38] Determining what people are thinking just by looking at their eyes can be highly challenging. It's not too difficult to tell whether they are happy or sad, but could you tell if they were suspicious or indecisive? Accusing or irritated? Such differences are detectable, but not everyone is adept at detecting them. People who read a lot of fiction turn out to be more adept.[39] They can distinguish flirtation from gratitude or caution from boredom just from the eyes. Contemplating emotions in fiction appears to improve the detection of these emotions in real life, even though we never see anyone's eyes when reading.

Of course, the association between fiction and mind reading could run in the opposite direction: people who are good at detecting subtle emotions may be drawn to fiction, where they will further explore these emotions, among other mental states. Is there any reason to think that fiction plays a causal role in social reasoning? Additional studies, employing an experimental design, suggest yes. People who are asked to read a passage of fiction show improvements in social reasoning from before the passage to after, particularly if they are absorbed by it.[40] These improvements include enhanced empathy, enhanced perspective taking, better emotion detection, and increased prosociality. Such improvements are not observed for people who spend the same amount of time reading nonfiction.

Reading about minds thus improves mind reading. From a developmental perspective, the interplay between these activities is grounded in pretense. Children engage with fictional minds long before they can read fictional stories, in the form of pretend play. They attribute minds to dolls and stuffed animals as well as beings that have no physical form: imaginary companions. For a young child, an imaginary companion provides opportunities for simulating social experiences much like those provided by fictional characters. And, like adults who read fiction, children who have imaginary companions exhibit advanced social reasoning.

For instance, preschoolers with imaginary companions outstrip their peers on their understanding of mental states like belief. They are more likely to understand that beliefs change, that beliefs can be false, and that people can hold different beliefs about the same situation.[41] Preschoolers with imaginary companions also have more advanced communication skills. They are better at gauging common ground in a conversation, and they tell better stories—stories with more dialogue, more descriptive language, and more action sequences.[42] These skills extend to private speech as well; children with imaginary companions use private speech to monitor and regulate their behavior more effectively than children without such companions.[43] Communicating with an imaginary companion can provide practice at communication in general, though a causal link between these factors has yet to be established.

When children create imaginary companions, they sometimes also create imaginary worlds—places where their companions live. These worlds, known as "paracosms," can be viewed as precursors to the full-fledged fictional worlds created by adults, specifically adult authors. One of every six children creates a paracosm at some point in development.[44] Paracosms vary in content from child to child, but they include, by definition, a unique landscape and a unique set of inhabitants. Some also include idiosyncratic details, like fountains that spray honey, cats that fly, and people who only read books they have written themselves.

Children's propensity to invent imaginary people and imaginary places would seem to contradict the earlier claim that they prefer realism to fantasy. Indeed, the research on children's imaginary companions describes many fantastic examples, including companions like Derek, a ninety-one-year-old man who is only two feet tall but can hit bears; Station Pheta, a boy with a big blue head who hunts for dinosaurs at the beach; and Yellow Granny, an invisible girl with a dyed ponytail who lives at the Tower of London.[45]

But for every Station Pheta or Yellow Granny, there are plenty of mundane companions: imaginary brothers who talk and play like real brothers, imaginary girls who look and dress like the girls at school, and imaginary dogs who bark and pant like the dog next door. Children may benefit from engaging with fictional people and places, similar to adults, but there's no evidence that the people and places they create themselves are particularly creative. In fact, researchers who have dissected the details of well-developed paracosms find them to be surprisingly mundane. The more time children spend developing their paracosms, the more they resemble the real world, with ordinary features like hospitals, post offices, and train stations. Paracosm building, in the words of one research team, "looks much more like work than play."[46]

Figure 9.3 Children's imaginary worlds are often surprisingly mundane. Here, a child depicts an imaginary world composed mainly of streets and railway stations. (Adapted from Cohen and MacKeith 1991) (Courtesy of the author)

Fiction, in general, can be quite mundane. Although there are lessons to be learned from simulating complicated minds in complicated situations, like those found in Faulkner or Tolstoy, most people prefer simpler scenarios: boy meets girl, hero rescues victim, detective finds killer, warrior vanquishes foe. Our hungry theory of mind is easily sated with a Hallmark rom-com or a Lifetime thriller. Indeed, an analysis of over 6,000 film scripts revealed that a large portion conform to the same plot, colloquially dubbed "man in a hole."[47] The protagonist in this plot (the man) encounters a setback (the hole) that he eventually overcomes. This plot is not only the most popular but also the most profitable, bringing in larger box office revenues than plots with more twists and turns.

The "man in a hole" plot has also been described as the "sympathetic" plot because audiences can sympathize with the protagonist as he attempts to reverse his ill-fortune.[48] The anthropologist Manvir Singh argues that a variant of this plot, in which the protagonist is a child and the ill-fortune is a villain or monster, is ubiquitous across cultures. He summarizes the plot as follows: "Once upon a time, a strong, attractive hero lost one or both of his parents. He then overcame a series of obstacles and faced off against a monster that had terrorized his community. The hero vanquished the monster and was celebrated." Singh notes that this is the story of Harry Potter, Luke Skywalker, Superman, and the Lion King. It's also the story of the Sotho hero Litaolane, the Garo hero Jereng, the Ainu hero Yayresu, and the New Guinean heroes Kototabe and Kelokelo.

People around the world crave formulaic plots. And the formulas they like best are those that confirm their beliefs rather than challenge them.[49] The characters in these narratives are also expected to be formulaic. They are expected not only to conform to stock roles like hero, villain, and victim but also to wear their thoughts and feelings on their sleeve. People in the real world are pretty good at hiding what they're thinking, but people in fiction are not.[50] They reveal these thoughts on their face, in their posture, through their actions, and in their speech. Fictional characters satisfy our craving for social stimulation without exhausting our social-reasoning abilities, the way real people often do.

The Moral of the Story

Fiction stimulates our social imagination in another way, beyond mind reading, by providing models of morality. Most fiction involves some kind of moral dilemma and some kind of solution to that dilemma. The solutions can be instructive, either as actions to imitate or transgressions to avoid. In many ways, it's the transgressions that are most instructive because we can observe their consequences without experiencing them ourselves. "Learning does not consist only of knowing what we must do," notes William of Baskerville, the protagonist from Umberto Eco's *The Name of the Rose,* "but also of knowing what we could do and perhaps should not."[51]

Of course, not all stories intended to teach moral lessons are adept at doing so. Fantastical content can compromise our ability to learn moral lessons, similar to how it compromises our ability to learn factual information, at least for children. One striking demonstration comes from a study where kindergarteners watched an episode of *Clifford the Big Red Dog* intended to teach children about being kind to individuals with disabilities.[52] In this episode, Clifford and his friends—who are all dogs—meet a new dog who is missing a leg. Some of Clifford's friends treat this dog as incompetent, offering him help he doesn't need. Another friend claims the dog's condition is contagious and tries to avoid him. In the end, Clifford and his friends learn that the three-legged dog wants to be treated like any other dog, and they all become friends.

Following the episode, children were asked what lesson the show was intended to teach. Only 19 percent were able to articulate a lesson about cooperation or tolerance. Most were unable to articulate any lesson at all. The children were then given a list of options to choose from, namely, (1) "if you do something wrong, you should always tell the truth, even if sometimes you'll get in trouble"; (2) "if you see a three-legged dog or cat or other animal, you should be nice to it"; and (3) "if you meet a kid who has a handicap, like they have to use crutches, you should still be friends with them." Only 11 percent picked the correct lesson (lesson 3). The rest picked either an irrelevant lesson (lesson 1) or an overly literal lesson (lesson 2), suggesting they had learned either no lesson or the idiosyncratic lesson that you should be kind to three-legged dogs.

Similar findings have been observed for stories involving other types of fantasy content and other moral lessons.[53] Children show little evidence of learning moral lessons, either in their speech or their behavior. They are not able to articulate the lesson, nor do they adjust their behavior in accordance with the lesson. For instance, the picture book *Little Raccoon Learns to Share* is intended to teach children how to share, as the title implies.[54] But researchers who read this book to young children found that it did not increase their willingness to share. In fact, many children shared less after the reading, possibly because they understood the lesson as pertaining specifically to raccoons.

The best way to teach children moral lessons, it turns out, is to tell them those lessons directly. Children are vastly more likely to change their attitudes and behaviors when given explicit instruction than when exposed to the same instruction in the form of a story.[55]

In one study, researchers attempted to change how six- and seven-year-old children share resources using stories about anthropomorphized beavers.[56] The children watched two workers complete a joint task, where one worked twice as hard as the other. They were then asked to give stickers to the workers. Some children rewarded the harder worker with more stickers, and others divided the stickers equally.

Next, children were told stories where the opposite strategy was endorsed. Children who divided resources equally were told a story about beavers who split the wood they gathered in proportion to how hard they worked, and children who divided resources by merit were told a story about beavers who split the wood equally, despite differences in labor. The beavers gave persuasive reasons for their decision, whatever that decision was, but these stories had no impact on children's allocation strategies, assessed again after the story. What did impact children's strategies was direct testimony from an adult, who told them that their preferred strategy was less fair than the alternative. Children not only changed their strategy but also persisted in using the alternative strategy several weeks later.

Fiction thus has limited potential for expanding children's moral imagination. Its potential for expanding adults' moral imagination is also limited. Although adults may be better at identifying the moral of a story, we have trouble imagining scenarios that contradict our moral intuitions, such as the intuition that sharing is good or the intuition that murder is wrong. Morality is expected to operate the same in fictional worlds as it does in the real world, just like physics and mathematics.[57] We are not just reluctant to imagine worlds in which morality operates differently; we find the task onerous, if not impossible. We experience what philosophers call "imaginative resistance."

For instance, can you imagine a world where killing baby girls is the right thing to do? How about a world where mammoths roam the streets of Los Angeles? Most people find the first world substantially more difficult to imagine than the second. Many also classify the first world as impossible.[58] Factually deviant worlds, like a world where mammoths still exist, are easy to entertain; morally deviant worlds are not.

In one study, the participants were asked not just to imagine a morally deviant world but to write a story about it—a story that included the statement "In killing her baby, Giselda did the right thing; after all, it was a girl."[59] The story was supposed to provide context for the statement so that readers might accept it as true. Participants also wrote stories around the statement "Packs of wolves were roaming the towns of England." The stories that participants wrote about infanticide tended to be much longer than the stories they wrote about wolves, yet many participants denied they had succeeded at making the statement about infanticide true. They were unable to construct a world where killing a baby girl is the right thing to do.

Take, for instance, this story set in a zombie apocalypse: "Giselda's husband, Dave, was very experienced in mixed martial arts, and did a remarkable job protecting her. She had never even been in a super dangerous encounter with a zombie since Dave protected her so well. Giselda discovered that she was expecting a baby. Dave and Giselda hoped with all they had that it was a boy, since baby girls were the most vulnerable. Unfortunately, Giselda had a baby girl who was missing her legs. They had no choice but to kill the baby. In killing her baby, Giselda did the right thing; after all, it was a girl." The author of this story denied that she had succeeded at justifying Giselda's actions, and others would likely agree. Giselda's motivation for killing her baby was that it had no legs, not that it was a girl.

Imaginative resistance can spill over to real events too, if real events are so morally deviant that we have trouble imaging how they could happen. People have publicly denied many moral atrocities, including the Holocaust, the Armenian genocide, the terrorist attacks of 9 / 11, the shooting at Sandy Hook Elementary School, and the January 6th invasion of the US Capitol.[60] As noted in chapter 2, when we hear of such atrocities, the first question we ask ourselves is often a question about possibility: "How could this happen?" Some people deal with this question by "rewriting" the event, creating a more palatable version they can accept. Others deny the event altogether. There are, of course, ideological motives behind denying an atrocity, when the denier wants to protect the perpetrators or avoid dealing with their crimes, but that ideology finds a foothold in imaginative resistance. "Truth can be stranger than fiction," observed Mark Twain, because "fiction is obliged to stick to [familiar] possibilities; truth isn't."[61]

In sum, the relationship between fiction and imagination is both mutually supportive and mutually constraining. Fiction is made possible by imagination, by decoupling fictional events from real events, but we still use the latter to reason about the former.

Fiction can help us discover new causal relations, new social relations, new problem solutions, and new moral obligations. Fiction can also help us use these ideas by simulating situations where they might apply. We can learn from *Dune,* for instance, that flouting a social norm can be used as a test of loyalty without risking the consequences of flouting real norms in real life.

On the other hand, learning from fiction is more the exception than the rule, especially among children. We enjoy fiction but only if it conforms to our expectations about real-world causation and only if it is processed easily by our hungry, yet lazy, theory of mind. Fantasy—the kind of fiction we most strongly associate with imagination—is an acquired taste. It is also a poor model for learning because we are inclined to quarantine the information encountered therein. What happens in fantasy stays in fantasy. While fiction can expand imagination by providing us with alternative models of reality, not all models are equally useful, and not all uses are equally judicious.

10 Religion

Expanding Our Metaphysical Imagination

Some of humanity's most imaginative ideas can be found in the context of religion. Ideas about gods, spirits, souls, and the afterlife defy perception, as well as expectation, and thus draw heavily on imagination. Consider some of the core tenets of Christian theology. The central figure in this theology is an all-knowing, all-powerful being that exists outside of space and time: God. God created the universe and now watches over every part of it because God is everywhere at once. God then populated this world by creating human beings, sculpted from earth and animated with an immortal soul. God also created another set of beings—angels—to act as conduits to humans. Angels, like God, are eternal and incorporeal. They appear to humans as visions or voices, and they guide humans' souls to an otherworldly realm—Heaven—upon the death of their physical bodies so they can reunite with God.

These entities—God, angels, and Heaven—have no physical form and do not therefore follow physical laws. Understanding their

Figure 10.1 Angels are ethereal beings in formal theology, but they are frequently depicted as humans with wings. This eighteenth-century etching depicts the angel that expelled Adam and Eve from Eden as so humanlike that he requires a cloud to stand on. Adam also appears to have a belly button. (Wikimedia Commons)

nature requires constructing a novel set of possibilities, distinct from the familiar possibilities of ordinary people and ordinary places, at least in principle. In practice, most Christians seem to understand God, angels, and Heaven in more concrete terms.

Walk through any cathedral or gallery of medieval art and you'll see God depicted as a human—typically an old man, clad in flowing white robes and adorned with a long white beard. Angels are also depicted as humans, though of variable age, gender, and dress. They are distinguished from God by their wings, which they presumably need for traveling between Heaven and Earth. When depicted in Heaven, they retain their human form, typically standing or sitting on clouds, as if they require support. Depictions of the first humans—Adam and Eve—also violate official theology because they typically have belly buttons. Beings sculpted from earth (or from Adam's rib, in Eve's case) would not have been connected to an umbilical cord and should therefore have smooth stomachs, unmarred by the vestiges of birth.

It's possible that divine beings are depicted as human beings in paintings and statues because there would be no way to depict their theologically correct form. How might an artist depict a being that has no body or a being that is everywhere at once? Artists are forced to depict divine beings concretely. But these depictions turn out to be accurate representations of how the average person imagines them. We may know the theological view of God, angels, and other divine beings, as codified in religious doctrine and disseminated through religious teachings, but we rarely reason about those beings in theologically correct ways.[1]

For instance, when asked to describe God, most people do not use terms like "infinite," "eternal," or "incorporeal" but rather "man," "beard," and "robe."[2] They describe God as "a man in white robes with long brown hair and tan skin," "a powerful man who watches over everybody," or "a big, bulky guy who is very nice." When asked to select one of several descriptions that best matches their conception of God, most select "God is a personal being and is involved in human affairs today." Far fewer select the theologically accurate description "God is not a person but is something like energy, the universe, or a cosmic force."

Findings like these suggest that artists' depictions of God as a superhuman, exemplified by Michelangelo's depiction on the ceiling of the Sistine Chapel, is viewed literally, not metaphorically. Most religious people imagine God in explicitly concrete terms. And not just God but other religious ideas as well, including angels, Satan, Heaven, and Hell. Research from my own laboratory has revealed that people's religious beliefs are surprisingly concrete—more physical than metaphysical.

For instance, people who hold anthropomorphic conceptions of God typically also hold anthropomorphic conceptions of angels and Satan, judging that all such beings engage in humanlike activities, including thinking, talking, growing, sleeping, sitting, and stretching. They describe angels as "people dressed in white with wings and halos" or "beautiful beings with pale skin, red cheeks, and flowing blonde hair," and they describe Satan as "an ominous figure, always surrounded by flames" or "like a human being that is all red and has a tail." Not everyone's views are this concrete, but concrete views are more common than abstract ones, even among people who profess strong belief in these beings.[3]

Heaven and Hell are also imagined concretely, as physical locations rather than spiritual destinations or states of being. Heaven is described as "white and spacious" with "lots of clouds and light," and Hell is described as "dark, dirty, and filled with fire," "deep in the center of the earth." The residents of Heaven are claimed to "live as they do among Earth, yet they are happier and more friendly," whereas the residents of Hell are claimed to "labor for the rest of eternity," forever "punished and tortured and living terrible lives."[4]

These beliefs, which seem to be grounded in our understanding of ordinary people and ordinary places, support concrete interpretations of other religious matters. People who think of God as a kind of superhuman tend to believe that God created humans in their current form, as opposed to creating the conditions under which humans evolved from nonhuman ancestors. They also tend to see human suffering as punishment directly inflicted on wrongdoers, as opposed to an inexplicable part of God's infinitely wise plan.[5]

Again, not everyone holds these beliefs, but they are a dominant pattern among those who do believe. For most believers, God is not an abstract entity—a kind of force, energy, or consciousness—but a person-like being who created humans in his own image and now watches over them, intervening in their affairs, judging their actions, and punishing their misdeeds. These ideas guide believers' behavior as well as their thoughts; the more they view God as person-like, the more they communicate with God through prayer and worship. A God who cares about people is presumably also a God who listens to people.

Such notions illustrate how religion, like other areas of thought, is characterized by a tension between ordinary, concrete ideas and extraordinary, abstract ones. The landscape of religious ideas includes omniscient beings, omnipresent powers, and imperceptible worlds, but these possibilities lie far beyond the familiar terrain of ordinary people, places, and events and are therefore difficult to contemplate, let alone accept as real. In this chapter, we will explore how knowledge of human affairs constrains our understanding of divine affairs, including how we imagine God, how we imagine the afterlife, how we decide which religious claims to believe, and how we apply those beliefs to the natural world.

Our understanding of divinity tends to be more secular than sublime, but we still exhibit marked changes in religious cognition over the course of development. Religious teachings provide alternative models of the universe and our place within it in much the same way that pretense and fiction provide alternative models of everyday life. Exposure to these models can expand the imagination, even if we do not view those models as literally true.

In God's Image

Religion is a cultural universal—practiced by most people in most societies—but the content of religious beliefs varies widely. The Christian notion of an omnipotent, omnipresent God is an anomaly in the full landscape of religious ideas. Most religions involve beings with less power and scope, such as local gods, nature spirits, or the ghosts of deceased ancestors.[6] Historically, nature spirits were the first recurrent feature of religion, followed by the afterlife, followed by shamanism and ancestor worship.[7] Belief in one almighty god did not emerge until the development of large-scale societies, where this belief fostered greater trust and cooperation than belief in different local gods.[8] A god who watches over everyone at all times is a better monitor of social behavior than gods who watch over particular people at particular times.

Today, most people believe in one almighty god, as most people are either Christian or Muslim, but this notion of divinity is conceptually novel. It contrasts with the more familiar notion of a person-like being, similar to ourselves but with special powers such as invisibility or immortality. Anthropomorphism, or the projection of human properties onto the world at large, has long been cited as the origin of religion.[9] This practice makes sense of animistic beliefs as well as belief in spirit ancestors and local gods. The best-known proponents of anthropomorphism are the nineteenth-century anthropologists Edward Tylor and James Frazer, but appeals to anthropomorphism date back to antiquity. The Greek philosopher Xenophanes of Colophon once noted, "If cattle or horses or lions had hands and could draw and could sculpt like men, then the horses would draw their gods like horses, and the cattle like cattle; and each they would shape bodies of gods in the likeness, each kind, of their own."[10]

Conceptions of divine beings thus fall on a continuum, from anthropomorphic to abstract. Anthropomorphic conceptions are ancient and intuitive, but abstract conceptions now dominate today's religious landscape, at least from the perspective of formal doctrine. Abstract conceptions take work to imagine and effort to maintain, so they remain in tension with anthropomorphic ones—a tension that can be seen across cultures, contexts, and development.

Culturally, divine beings are depicted and described in vastly different ways. Consider the differences between Islamic, Hindu, and Judeo-Christian cultures. In Islamic culture, iconography and anthropomorphism are actively prohibited. Beings like Muhammad and Allah are not depicted in paintings or statues, and their depictions in text emphasize their abstract, nonhuman properties.[11] In Hindu culture, on the other hand, anthropomorphic representations are explicitly encouraged. Beings like Ganesha and Krishna have human bodies, carry human artifacts, perform human activities, and engage in physical interactions with humans and other gods. Judeo-Christian culture occupies more of a middle ground. Depictions of God range from the highly abstract, corresponding to descriptions like "first cause," "unmoved mover," and "universal spirit" to the highly anthropomorphic, corresponding to descriptions like "heavenly father," "divine ruler," and "intelligent designer." A Google search of "God" in the United States will bring up images of an old man alongside images of light, energy, and the universe.

The particular mix of anthropomorphic and abstract representations in one's culture influences how people imagine God. When asked whether God possesses human properties like the ability to talk, grow, or move, Muslim adults attribute few such properties, whereas Hindu adults attribute most. Judeo-Christian adults fall somewhere in between.[12] This variation is most pronounced for body-dependent properties like eating, sleeping, and sitting and is least pronounced for mind-dependent properties like thinking, feeling, and knowing. People across cultures disagree about the form God takes, but they generally agree that God has a mind.

Developmentally, this distinction between body-dependent and mind-dependent properties does not emerge until children have had ample exposure to religious concepts because children's earliest conceptions of God are anthropomorphic in both mind and body.

They tend to attribute *all* human properties to God, regardless of cultural upbringing or family environment.[13]

In one of my own studies, we asked five-year-olds from Judeo-Christian households to decide whether God possesses a variety of human properties, as well as to describe God in their own words.[14] We then compared children's responses to the responses of their parents, who answered the same questions in another room. Children's responses converged with their parents' when it came to mind-dependent properties: everyone agreed that God thinks, dreams, and talks. But when it came to body-dependent properties, their responses diverged. Children were twice as likely as their parents to claim that God eats, sneezes, and grows (biological properties) and that God sits, stretches, and jumps (physical properties).

Children were also more likely to describe God in explicitly anthropomorphic terms. One child described God as "a person that makes stuff," whereas that child's parent described God as "the source of the universe, the endless being." Another child described God as "a person that ruled the whole world once, even the fish," whereas their parent described God as "the spiritual presence in all things; that which inspires us to be good."

Five-year-olds know that God is not a human; they acknowledge that God cannot be seen, that God can be in multiple places at the same time, and that God knows things ordinary humans do not.[15] But five-year-olds append these extraordinary properties to the concept of an ordinary human. It takes additional instruction or reflection for children to reimagine God as something less anthropomorphic and more abstract. And not everyone does reimagine God in this manner. Some adults retain anthropomorphic views of God throughout their lives, as noted previously, and some cultures support anthropomorphic views, obviating the need to rethink them.

This interaction between culture and development is well illustrated by a study that my colleagues and I conducted in India. We investigated how Hindu and Muslim children conceptualize divine beings, both those within their own religion and those within the other religion.[16] Hinduism is the dominant religion in India, but Islam is common as well, which means that children growing up in India learn about Hindu beings, like Ganesha and Krishna, as well as Islamic beings, like Muhammad and Allah. The former are described and depicted as having humanlike bodies, whereas the latter are described abstractly and depicted infrequently. The juxtaposition of such discrepant representations raises questions about how children conceptualize the two types of beings and whether their religious upbringing, as Hindu or Muslim, influences that conceptualization.

Consistent with findings from Judeo-Christian populations, we found that younger children anthropomorphized divine beings more strongly than older children. But the nature of the being mattered, as did the children's religion. Hindu beings were anthropomorphized by children of all ages and both religions, but Islamic beings were anthropomorphized more consistently by younger children and by Hindu children.

As an illustration, consider how young Hindu children conceptualized Krishna (a Hindu being) and Allah (an Islamic being) relative to older Muslim children. Both groups agreed that Krishna possesses many human properties, including the ability to eat, grow, jump, and sit, but they disagreed about whether Allah possesses these properties. Young Hindu children attributed nearly as many human properties to Allah as to Krishna, but older Muslim children attributed far fewer to Allah. If they attributed any, they attributed mind-dependent properties, such as the ability to think and talk, rather than body-dependent ones, such as the ability to jump and sit.

Findings like these suggest that when we imagine divine beings, we default to our conception of human beings, and we stick with that conception unless prompted to reimagine those beings. Many cultures provide such prompts, in the form of abstract depictions of divinity in art, literature, or discourse, but even people who come to embrace fully abstract concepts continue to hold onto their earlier, anthropomorphic ones. The two concepts coexist, creating conflict when reasoning about God's nonhuman properties, such as omniscience and omnipresence.

Some of the earliest evidence for coexisting God concepts came from a study where Judeo-Christian adults were asked to read stories about God and then recall those stories from memory.[17] Nearly all participants claimed that God is omniscient and omnipresent when asked directly, but they recalled the stories in ways that implied that God is limited in what he knows and where he can be. One story, for instance, described God listening to two birds at an airport in the midst of a jet landing. The story made no mention of limitations on God's attention or perception, but participants introduced such limitations in their recall, claiming that the jet "took God's attention away" or that "God could only hear the jet" and "could no longer hear the birds."

These findings suggest that people hold two god concepts: an abstract one, used when reasoning about God in theological contexts, and an anthropomorphic one, used when reasoning about God in everyday contexts like stories. But a problem with this interpretation is that the stories themselves may have triggered participants' anthropomorphic inferences because these stories described God with a heavy dose of anthropomorphism, including statements like "God was aware of the girl's deed and was pleased by it" and "God was helping an angel work on a crossword puzzle."

Better evidence for coexisting god concepts comes from statement-verification tasks, where participants are shown a series of statements about God and asked to judge each as true or false as quickly as possible.[18] Some statements are consistent with an anthropomorphic concept of God and others are consistent with only an abstract concept, and participants find the latter reliably more difficult to verify. They take longer to verify such statements and also make more errors.

As an example, consider the statements "God can occupy the space inside a church" and "God can occupy the space inside a boulder." Both are consistent with an abstract, theologically correct concept, but only the first is consistent with an anthropomorphic concept; the second actively conflicts with such a concept. This conflict should not interfere with participants' reasoning if they hold only one god concept, having replaced the anthropomorphic concept developed in childhood with an abstract concept developed later in life, but it does interfere. Participants verify statements that violate the core properties of humans, like "God can hear what I say to myself," more slowly and less accurately than statements that accord with those properties, like "God can hear what I say out loud." This pattern has been observed in adults of all ages, from eighteen to eighty, and for a variety of divine beings, including Jesus and the Holy Spirit.[19]

Our imagination of divine beings is thus constrained by our knowledge of human beings. The first God concepts we form are anthropomorphic, and we retain those concepts even as we embrace abstract alternatives like "cosmic energy," "infinite force," and "universal consciousness." While the majority of religious people endorse explicitly anthropomorphic concepts throughout their life, either because their culture supports such concepts or because they find such concepts personally appealing, the minority who come to endorse an abstract concept still think of God as a superhuman, at least implicitly.

The Mind of God

In February 2022, a Catholic priest in Phoenix, Arizona, resigned as pastor of his parish after church officials discovered that he had performed thousands of baptisms incorrectly. His mistake had nothing to do with the actions he made or the instruments he used; it was the words he said: "We baptize you in the name of the Father and of the Son and of the Holy Spirit." The correct incantation is "*I* baptize you in the name of the Father and of the Son and of the Holy Spirit," and the difference between "we" and "I" appears to matter a lot in the eyes of the church—or its ears, as the case may be.[20]

Baptism is a Catholic requirement for salvation, and salvation is a requirement for entry into Heaven, so any deviation from the established ritual could have eternal consequences. On the other hand, Catholic doctrine stipulates that God is omniscient, and an omniscient being should be able to recognize the intention behind the priest's incantation. Does God care whether "we" is substituted for "I"? The answer to this question depends on how we understand God's mind. Theological conceptions of God, in the Judeo-Christian tradition, place no boundary on God's knowledge or wisdom, but this conception defies our understanding of ordinary minds as finite and fallible. An infinite and infallible mind stretches imagination.

Our earliest understanding of God's mind appears to be fully anthropomorphic.[21] Four-year-olds think that God's knowledge is limited, similar to humans, and that God can be uninformed or even misinformed. For instance, when four-year-olds are shown a closed box and asked whether God would know what's inside without opening it, they claim that God would not. When shown a box with a misleading appearance, like a cracker box that had been emptied of crackers and filled instead with rocks, they claim that God would think the box contains crackers, not rocks.[22]

By early elementary school, children recognize that God possesses knowledge that humans do not, but they continue to treat God's mind as limited. In one study, elementary-school-aged children were reminded that an omniscient being like God knows everything about everything.[23] They were then asked who would know more about how to fix a car or how to treat a runny nose: the omniscient being or a human expert. Children typically sided with the expert, claiming that a car mechanic would know more about how to fix a car and a doctor would know more about how to treat a runny nose. Children of this age also have difficulty appreciating that an omniscient being is aware of everything, including all prayers. They think that prayers said out loud are more likely to attract God's attention than prayers said in one's head, particularly if those prayers are said at church.[24]

Of course, many adults prefer to pray out loud at church than silently at home.[25] For all our practice thinking of God as omniscient, we still exhibit signs of construing God's mind as limited and fallible.[26] One sign, as noted earlier, is that we impose limits on God's awareness and perception when reasoning about God in the context of a narrative or when verifying statements about God under time pressure. Another sign is that when we pray to God we recruit regions of the brain involved in reasoning about the minds of ordinary humans, namely, the temporoparietal junction and the medial prefrontal cortex.[27] These regions are active when we pray but are not particularly active when we recite incantations, like the Lord's Prayer, suggesting that we engage with God's mind *as a mind* and not just a symbol or an abstraction.

Yet another sign that adults treat God's mind as a human mind comes from studies of what we think God cares most about—what God is paying attention to. Participants in these studies are asked to decide, as quickly as possible, whether God knows a variety of facts.[28] Some are about the natural world ("Does God know the structure of plant DNA?") and some are about people ("Does God know how fast Joey's heart beats?"). Among the facts about people, some are about people's good behavior ("Does God know that Ann gives to the homeless?"), and some are about people's bad behavior ("Does God know that John cheats on his taxes?"). While participants typically affirm that God knows all these things, they affirm God's knowledge of people more quickly than his knowledge of other facts, and they affirm his knowledge about people's bad behavior the quickest. God may know everything, but it's his social knowledge that preoccupies our attention, particularly his knowledge of social transgressions. God, like Santa, is keeping track of whether we've been naughty or nice.

This egocentric perspective of what God knows extends to our perception of what God thinks is morally right. Does God support same-sex marriage? Does God support affirmative action? Our answers vary with whether we ourselves support same-sex marriage and affirmative action. The moral beliefs we attribute to God resemble our own beliefs more than the beliefs we attribute to other people, including people with similar political orientations.[29] The same has been found for our perception of Jesus's beliefs, if Jesus were alive today.[30] Liberal Christians claim that Jesus would support same-sex marriage and affirmative action, whereas conservative Christians claim he would oppose them. Our conviction that divine beings believe the same things we do suggests that we use our mind as a model for theirs. We bypass the difficulty of imagining an infinite mind by tapping into the finite workings of our own.

Life after Death

Religions are traditionally organized around belief in divine beings, but many also posit ideas about the afterlife—a state of existence beyond death. Is it possible to imagine such a state? Religious doctrines describe the afterlife as a place for souls, not bodies. Souls are thought to be the immaterial part of identity and consciousness, which survives the death of the physical body and migrates to a nonphysical realm of existence, such as Heaven or Hell.

Consistent with these ideas, adults raised in Judeo-Christian societies typically claim that when people die they retain their mental states but not their bodily states. For instance, in one study, Judeo-Christian adults were told about Richard, a middle-aged teacher who died in a car accident on his way to work. They were asked whether Richard, now dead, still possesses mental states like thoughts, feelings, and desires as well as bodily states like hunger, thirst, and nausea.[31] The participants rarely claimed that Richard continues to possess bodily states, but they frequently claimed that he continues to possess mental states. And, on the rare occasions when participants denied the continuity of a mental state, it took them longer to do so than to deny the continuity of a bodily state, implying that these participants intuitively thought of Richard as a mind without a body and required extra time to override this intuition.

Judeo-Christian adults thus appear to imagine death as a spiritual transformation rather than a biological termination.[32] Some scholars have argued that humans in general imagine death in this way.[33] But findings from several lines of research suggest that the idea of life after death defies imagination. Death is most naturally understood as the cessation of *all* functions, bodily and mental. The possibility that minds survive bodies, in the form of souls, is a conceptual novelty, endorsed by some cultures but not others and by adults but not children. It's a possibility we learn from religious teachings but do not intuit on our own.

From a global perspective, many cultures treat death exactly as it appears: the cessation of all activity. When anthropologists have posed questions about life after death to people raised in non-Christian societies, such as the Marajó people of Brazil, the Vezo people of Madagascar, or the Shuar people of Ecuador, they all tend to claim that mental states cease with death just like bodily states.[34] People raised in secular societies, like China, also claim that mental states cease with death.[35]

In societies where people do believe that minds survive bodies, this belief does not emerge until late childhood or early adolescence.[36] It emerges in two stages. First, children come to understand death as the end point of life—as an inevitable and irreversible breakdown of the body. This understanding is achieved around age seven and yields the expectation that all functions cease with death, including mental functions. It's not for another five years that children reconcile this expectation with religious teachings about immortal souls and their eternal destinations, which prompt children to differentiate mental states from bodily states when reasoning about the deceased.

This differentiation is tenuous, though, and tends to fluctuate with context.[37] Children are most likely to differentiate mental states from bodily states in religious contexts, such as when death occurs in the presence of a priest and the deceased is described as "with God." They are least likely to do so in secular contexts, such as when death occurs in the presence of doctors and the deceased is described as "dead and buried." In the latter contexts, children usually default to a materialistic understanding of death and claim that all functions cease.

Context effects have been observed across ages and cultures.[38] They imply that it takes effort to maintain a dualistic conception of death, where bodies perish but minds persist. What, after all, might it mean for the deceased to exist without a body, as a formless soul in a spaceless afterlife? We can entertain this possibility in the abstract but have trouble embracing it in practice. Just think of the many movies and television shows that include characters who have died, such as *Ghost, Beetlejuice, The Good Place, The Sixth Sense, Casper, Coco,* and *Soul.* These characters have bodies—often the same bodies as when they were alive—and we do not object. On the contrary, we find embodied representations of the deceased perfectly plausible.

Such representations are typical in art and literature, both now and in premodern times.[39] In the *Odyssey,* for instance, Odysseus interacts with the dead as if they were still flesh and blood. Odysseus

Figure 10.2 The deceased, in many theologies, live on as immaterial souls, but depictions of the deceased typically include bodies. In this painting by Henry Fuseli, the deceased prophet Tiresias is depicted as having the same body and wearing the same clothes as when he was alive. Even Tiresias's blindness remains unchanged. (Wikimedia Commons)

travels to Hades, where the dead physically reside. He makes a sacrificial offering, which they bodily consume. And he speaks with the prophet Teiresias, who retains the same clothes and appearance as when he was alive. Similar encounters with the dead are described in the oldest piece of literature, the *Epic of Gilgamesh,* as well as the most popular piece of literature, the Bible.

Even religions that posit reincarnation, such as Hinduism and Buddhism, rely on embodied reasoning. Reincarnation would seem to require a strict division between minds and bodies because it stipulates that minds move from one body to another. Yet people who subscribe to reincarnation use physical features, like birthmarks and scars, to infer shared identity across bodies.[40] Bodily features are viewed as stronger indicators of past lives than psychological features such as shared behaviors or shared personalities. Even Western adults who do not believe in reincarnation instinctively focus on bodily features when reasoning about hypothetical cases of reincarnation, such as when identifying who a person might have been in their former life.[41]

Religious teachings predicated on the separability of minds from bodies defy imagination because all our experience with minds—either our own or other people's—occurs in the context of bodies. These teachings cover not only the afterlife but also divine beings, as discussed earlier. From a theological point of view, divine beings are disembodied agents, but we intuitively imagine them as having bodies, just as the afterlife is a place for disembodied souls but we intuitively imagine the deceased as retaining their bodies. Religion invites us to imagine minds as separable from bodies, but everyday experience ties them closely together.[42]

Why does religion make such demands on imagination? Why insist that gods and ghosts have no bodies when we are inclined to think they do? One possibility is that disembodied agents may help make sense of religious claims that contradict everyday observation. The claim that divine beings are watching us and monitoring our social interactions is difficult to reconcile with the fact that we never see them. This tension can be resolved by stipulating that divine beings have minds but not bodies, allowing them to see us without us seeing them. Likewise, the claim that there is life after death is difficult to reconcile with the fact that dead bodies rot and decompose. This tension can be resolved by stipulating that only souls transition to the afterlife, allowing for the continued existence of a person's consciousness and identity despite the end of their physical existence. Religious concepts may thus expand our sense of existence by challenging the embodied aspects of everyday life.

True Believers

In chapters 8 and 9 we discussed how the ideas contained in pretense and fiction provide models for exploring alternative versions of reality. Religious ideas also provide models, but they differ from pretend ideas and fictional ideas in that believers accept them as true. For a believer, religious ideas are not alternatives to reality but clarifications—insights into reality's true nature. Believing in religious ideas is not simple, though. Believers must tackle two challenges: the conceptual challenge of imagining disembodied agents, omniscient minds, and immortal souls, as well as the epistemic challenge of accepting these ideas as true despite a lack of perceptible evidence.

Some scholars have argued that religious beliefs are no different in kind from ordinary, matter-of-fact beliefs.[43] They claim that believers see the world as containing gods, spirits, and souls just as it contains rocks, rivers, and trees; the existence of divine beings is no more mysterious or uncertain than the existence of ordinary objects. This argument helps make sense of how thoroughly believers incorporate religious ideas into their lives—in the form of prayer, worship, and devotion—as well as the many sacrifices they make to uphold religious traditions. Why would believers devote so much time and energy to deities that may not exist?

The effort believers put into their beliefs would seem to be proof of their conviction, but the anthropologist Tanya Luhrmann thinks the connection between effort and belief works in reverse: that believers are convinced of their beliefs *because of* the effort they put into them.[44] Religious practices, in Luhrmann's view, are not by-products of belief but means of achieving it, by turning vague abstractions into palpable experiences. Religious ideas become real by connecting them to people and places whose reality is uncontested. This activity requires a different mindset than matter-of-fact beliefs—a mindset Luhrmann calls the "faith frame." Within the faith frame, we engage with religious claims as true while simultaneously recognizing that their truth depends on our engagement, similar to how pretense becomes real when everyone involved in the pretense treats it as real. The faith frame makes religious claims subjectively true whether or not they are objectively true as well.

Several findings support the idea that religious beliefs are qualitatively different from ordinary beliefs and thus require a special mindset or attitude.[45] First, believers are reliably less confident in their religious beliefs than in their ordinary beliefs, including beliefs about other unobservable entities, like X-rays, electrons, and black holes. Scientific entities defy observation and intuition just like religious ones, but people are significantly more confident in the existence of scientific entities.[46] This difference emerges in the preschool years before children have had much instruction in either religion or science, and it emerges in cultures with strong religious traditions, like Spain and Iran, no differently than in more secular cultures, like the United States and China.[47]

Parallel to these differences in confidence, believers exhibit reliable differences in their reality judgments; they take longer to classify religious entities as real relative to scientific ones.[48] That is, when believers are shown words on a computer screen and asked to press a certain key when the word denotes something real, they take longer to press the key when shown "angel" or "soul" than when shown "electron" or "black hole."

Religious beliefs are also less responsive to evidence than are ordinary beliefs. A belief like "the earth goes around the sun" may be accepted on faith when first acquired, but most people expect that there is evidence supporting this belief and that we can learn the evidence if desired. We might even change our mind if we are not convinced by the evidence. A belief like "human beings have souls," on the other hand, does not call for evidence in the same way. We are willing to accept this belief as a mystery, both for ourselves and for others, and we are not particularly bothered by the possibility that no evidence might be found.[49]

The disconnect between religious beliefs and evidence can also be seen in our openness to religious relativism.[50] In factual disputes, such as a dispute over whether Benjamin Franklin was once president of the United States, we claim that people on opposite sides of the dispute cannot both be right. Someone is right, and someone is wrong. But we are more tolerant of opposing beliefs in religious disputes, such as a dispute over whether there is life after death. While we may have a preferred position ourselves, we are willing to grant that people who hold the opposite position may also be right, at least for beliefs that are not central to our identity.

Finally, we talk about religious beliefs and ordinary beliefs in distinct ways. When we ascribe ordinary beliefs to other people, we use the verb "think," as in "Jane thinks Benjamin Franklin was president of the United States." But when we ascribe religious beliefs, we use the word "believe," as in "Jane believes she will go to Heaven when she dies." In the Corpus of Contemporary American English—a massive collection of English-language documents—the word "believe" regularly co-occurs with religious words, but the word "think" does not.[51] There are 113 words that regularly co-occur with "believe," and fourteen of them are religious: God, miracles, reincarnation, afterlife, redemption, sanctity, Allah, Messiah, Creationism, devoutly, exorcism, transubstantiation, reincarnated, and Anti-Christ. None of these words regularly co-occur with "think."

This pattern has been observed in behavioral studies as well. When English speakers are asked to complete a sentence like "Jane ___ that Jesus Christ died for human sins," they typically write "believes," whereas they write "thinks" when completing a sentence like "Jane ___ that bronze contains more copper than tin."[52] The same is true for Fante speakers in Ghana, Thai speakers in Thailand, Mandarin speakers in China, and Bislama speakers in Vanuatu.[53] All use a verb that corresponds to "believe" in the context of religious ideas but use a verb that corresponds to "think" in the context of ordinary, matter-of-fact ones.

The special language we use to discuss religious ideas signifies to our conversational partners that we deem these ideas epistemically unique. This language may also have significance for children because it marks religious beliefs as qualitatively different from ordinary ones. Children do not have to inspect Gallup polls or Pew surveys to realize that religious beliefs are held with less certainty and shared less widely. They can infer as much from how we talk about them.[54]

Psychologists have explored the relation between language and belief by introducing children to novel entities, like "cusks," in different linguistic contexts. When children hear about "cusks" through statements of belief, as in "I believe in cusks" or "cusks are real," children remain skeptical of cusks' existence. But when they hear about cusks through statements of fact, such as "I saw some cusks" or "I studied cusks," children are more likely to accept that they exist.[55] These findings suggest that, while the mindset underlying religious beliefs can be profoundly different from that underlying ordinary beliefs, that difference can be established quite easily with a simple turn of phrase. Statements like "I believe in angels" and "I believe in souls" render their referents less believable, not more. It takes commitment and devotion to imagine angels and souls existing on the same plane as electrons and black holes.

The Hand of God

The Greek historian Thucydides is often credited as the first historian who explained human affairs without reference to divine intervention. Historians before Thucydides often appealed to gods and their dispositions, such as Hermes and his trickery or Hera and her jealousy, but Thucydides limited his explanations to human agency.[56] This decision sounds as if it may have been effortful, given that Thucydides was raised to believe in gods and was writing for an audience that believed in them as well. But empirical studies suggest that the opposite is effortful: explaining events in terms of divine intervention takes work, even for devout believers. Divine intervention is not a possibility that intuitively springs to mind.

Consider aspects of the natural world you may not fully understand, such as how bones heal or how clouds produce rain. God provides a ready means of answering these questions: bones heal because God wants us to recover from injuries, and clouds produce rain because God wants plants and animals to have something to drink. Yet appeals to God are rare.[57] Instead, we appeal to natural processes like cell growth and the water cycle—the kinds of causal mechanisms discussed in chapter 5. Even devout believers prefer natural explanations to supernatural ones. In one study, researchers asked both Christians and atheists to explain natural phenomena and then submitted those explanations to an algorithm that looked for differences between them. But the algorithm failed; it could not find differences between the explanations provided by Christians and those provided by atheists because both groups appealed to the same kinds of causal mechanisms.[58]

Similar findings have been observed for how people of various ages and religions explain personal events, like recovering from an illness or getting a divorce.[59] Our first inclination is to identify natural causes, even when the events seem incredibly auspicious. Consider the following story: "About two years ago, Veronica became very sick with cancer. Because of the cancer, Veronica had to quit her job, so she did not have enough money to buy medicine. Veronica became very sad because she thought she would never get better. Then one day Veronica woke up and did not feel sick anymore. She went to the doctor, and the doctor told her that her cancer had gone away. Veronica asked the doctor, 'How did the cancer go away?' The doctor said that he did not know."

This story was written specifically to evoke thoughts of divine intervention, yet only a third of adults mentioned God when asked what might have caused Veronica's cancer to go away, and even fewer children mentioned God. Those who did mention God mentioned natural causes as well, citing factors like diet and lifestyle alongside divine intervention.

Just as God is not at the forefront of our minds when explaining natural events, God is not at the forefront of our minds when predicting those events either. Studies have looked at whether children think extraordinary events are more likely if God is involved, and they find that God's involvement is not viewed as sufficient.[60] Children focus instead on whether they can identify natural causes. For instance, children in one study were asked whether the following event could happen in real life: "Stephanie and her friends were sailing a boat and got caught in a bad storm. Lightning flashed, and Stephanie fell out of the boat. Stephanie started to sink. But she prayed to God that she could walk on water. She walked on water back to the boat. The storm passed, and everyone was safe." Other participants were told a similar story except the part about God was replaced with "But just at that moment, a large whale came by. So Stephanie stood on the back of the whale who carried her back to the boat." Children were skeptical of the possibility of both events, but they were more skeptical of the version that involved God.[61]

Children are not uniformly skeptical of divine intervention; those raised in religious households are more likely to endorse this possibility than children raised in secular households. But even children from religious households tend to deny that God could make the impossible possible.[62] They deny that God has the power to violate causal regularities, and they almost never mention God when brainstorming ways of bringing about such events.[63]

One of my favorite demonstrations of this skepticism comes from a study conducted in Columbia.[64] The participants were kindergarteners from Catholic families attending a Catholic school. Their school was filled with religious iconography, such as crosses, shrines, pictures of Jesus, and statues of the Virgin Mary, and their daily instruction included Bible stories, lessons about God, and prayer. The researchers involved in this study suspected that children raised in this environment would be more accepting of the possibility of impossible events, especially when reminded of God's power. They began their study by telling children, "Remember, some things can really happen and some things are just impossible—they can never happen. But God can make some things happen even if they are impossible." They then asked the children about the possibility of several impossible events, including a tree growing money, a puppy turning into a cow, and a bear flying through the sky.

Most children claimed that none of these events are possible. And when asked to justify their judgments, some explicitly denied the suggestion that God could make them happen. "God cannot make a bear fly," explained one child, "because it is impossible." Another asserted that "God cannot do any of that."

Religious ideas thus have limited scope. From early in development, we privilege natural causes over supernatural ones and rarely ponder the possibility of divine intervention. We quarantine religious beliefs from ordinary ones, treating them as less certain and less warranted, and we have difficulty wrapping our minds around religious notions like omniscience, omnipresence, and immorality. Religions around the world posit models of the universe that expand our sense of reality and our place within it, but these models lie so far from ordinary experience that we have trouble understanding them, let alone accepting them as true. Religion may be widespread, but the capacity to engage with religious ideas, in their full metaphysical glory, requires an expansion of imagination.

11 Reimagining Imagination

In the 1987 movie *The Princess Bride,* the mercenary Vizzini hatches a plot to kidnap a princess and frame a rival kingdom for her disappearance. His plan is foiled at every turn. Vizzini and his crew think they've absconded with the princess undetected but soon discover they are being followed. They think they can outpace their pursuer but discover he is gaining on them. They think they can kill the pursuer by cutting his rope as he scales a cliff but discover he has clung to the cliff face. They think they can defeat him in a swordfight but discover he is a better fighter. At each foil, Vizzini shouts, "Inconceivable!" After the third "inconceivable," a fellow kidnapper, Inigo Montoya, quips, "You keep using that word. I do not think it means what you think it means."

This quip is one of the movie's most memorable lines. It captures the audience's amusement at Vizzini's failure of imagination. Vizzini's blunders are conceivable; he just hasn't put in the effort to conceive of them. This sense of inconceivability resonates with the many failures of imagination explored in this book. When children claim that it is impossible for a person to find an alligator under the bed, they have failed to imagine events that are fully within the realm of conceivability. Children know what alligators and beds are, and they can construct a mental image of an alligator under a bed, but they fail to conceive of circumstances by which an alligator might find its way under a bed in real life. Children's expectations about alligators—that they are large and live in swamps, far from people's houses—prevent them from recognizing that an alligator under a bed does not violate any physical laws and should therefore be possible.

Or consider children's failure to imagine how a pipe cleaner could be used to retrieve a toy at the bottom of a vertical tube. They know that pipe cleaners can be bent into different shapes, and they see that the toy has affordances that could be snagged with a hook, but they do not conceive of the pipe cleaner as a hook. Children can use a premade hook to retrieve the toy, and they can make their own hook if they have watched someone else make one, but they do not envision the possibility of a hook on their own.

Children are not alone in succumbing to an illusion of inconceivability; adults do so as well. We may recognize that improbable or unfamiliar events are possible, but we overlook these events when solving problems or making decisions, similar to how Vizzini overlooked the possibility that the man following him was an excellent sailor, climber, and swordsman.

We overlook the possibility that our tried-and-true strategy for playing chess is not always optimal and our tried-and-true strategy for calculating price reductions is not always applicable. We fail to see new ways of analyzing common experiences, such as analyzing traffic jams as a case of emergence rather than incompetence or analyzing a fixed tax as inequitable rather than neutral. We struggle to imagine hypothetical situations that might justify extraordinary actions and struggle to devise theoretical models that might account for extraordinary observations. We default to our knowledge of ordinary animals when asked to draw extraterrestrial ones and default to our knowledge of ordinary minds when asked to contemplate omniscient ones.

In this final chapter, we will revisit the many failures of imagination encountered in the previous chapters as well as the means by which we overcome them. The failures are invariably tied to knowledge, but so are the solutions. Knowledge determines the boundaries of imagination, keeping us within a familiar terrain of possibilities but sometimes pointing to new terrains. The relation between knowledge and imagination has been emphasized in every chapter, but we will end by considering how this relation undermines popular views of imagination and how it turns popular advice for improving imagination on its head.

Imagination in the Popular Imagination

The story of imagination that prevails today is a story of decline. Imagination is viewed as an innate capacity that fades or shrinks with age, similar to the loss of hearing or vision. This view stems from a larger picture of developmental decline, in which children are thought to possess an inherent brilliance or cleverness that we ruin as we indoctrinate them to the conventions of adult life. Consider this proclamation from pop psychologist Tony Buzan: "Very young children use 98 percent of all thinking tools. By the time they're 12, they use about 75 percent. By the time they're teenagers, they're down to 50 percent, by the time they're in university it's less than 25 percent, and it's less than 15 percent by the time they're in industry."[1] Buzan's numbers are pure fabrication, but they resonate with the popular idea that the core of human ingenuity lies in the minds of the young.

Children are certainly born with the capacity to imagine—to consider possibilities they have not physically experienced but can mentally experience nonetheless. If this capacity were not innate, it would be exceedingly difficult to acquire, and no one (to my knowledge) has proposed a model of how we might do so. Still, a capacity to imagine unexperienced possibilities is not the same as a predisposition to imagine conceptually extraordinary possibilities. Imagination evolved for the purpose of contemplating close counterfactuals, not far ones. Far counterfactuals require work. They require comparing and combining disparate ideas, as well as the more foundational work of acquiring the knowledge needed to support those ideas.

Throughout this book, we've seen examples of how children have not yet done that work. They perform poorly on tasks that require generating novel possibilities or entertaining alternative realities, contradicting the idea that they are wellsprings of imagination. This disconnect between children's imagination and popular perceptions of their imagination can be seen throughout the research we've reviewed on children's early cognition.

In chapter 2, we encountered this disconnect in children's judgments of what can happen in the real world and what cannot. Contrary to popular belief, children do not think everything is possible. They think only things that hew close to their limited database of experience are possible. Events that defy children's expectations are pronounced impossible, including events that defy relatively shallow expectations, like making pickle-flavored ice cream or painting polka dots on an airplane. Children are also skeptical of events they see depicted on television or in books, contrary to the popular belief that they believe everything is real. Children's skepticism is so profound that they fail to learn even true facts from popular media, such as the customs of an unfamiliar culture or the vocabulary of an unfamiliar language.

One aspect of possibility where children exhibit credulity is their willingness to believe in magical beings like Santa Claus and the Tooth Fairy. But even here, their belief is driven not by an openness to novel ideas but a willingness to accept the testimony of credible informants, such as parents and teachers. Children stop believing in magical beings when their understanding of physical possibility eclipses that willingness—when they can no longer suspend disbelief.

In chapter 3, we confronted the popular belief that children are natural-born innovators with studies documenting a striking absence of innovation in early childhood. Young children fail to bend pipe cleaners into hooks, twist yarn into lassos, or turn sticks into levers. Children are adept imitators, skillfully modeling the tools and techniques of others, but they fail to come up with their own innovations. In fact, children are so adept at imitating that they will imitate actions that have no bearing on the task at hand, such as twisting and prodding a puzzle box in ways that are patently unnecessary for retrieving the prize within.

This tendency to *over*imitate casts further doubt on the claim that children are natural-born innovators because any competent innovator should be able to identify unnecessary actions. The one setting where children do innovate is a group setting, where children can observe the actions of others. In such settings, children who stumble upon useful ideas will trigger a chain of copycats, who, in repeating the idea, will sometimes stumble upon an even more useful version. But groupwide innovations speak to the cleverness of evolution, not the cleverness of children. Even among adults, innovations almost always arise from the copying and tinkering of many individuals rather than the prophetic insights of just one.

Chapter 4 explored the idea that children are highly open-minded and thus receptive to new discoveries. But children embrace discoveries only if they accord with prior expectations, similar to adults. Anomalies are either ignored or rejected, both when evaluating observations made by someone else and when making one's own observations. If children expect objects to balance at their geometric center (rather than their center of mass), they will fixate on objects that do and ignore those that don't. If forced to confront counterexamples, children will concoct ad hoc explanations that allow them to retain their initial expectation. Children also have difficulty entertaining multiple possibilities at the same time, such as that some objects balance at their geometric center but others do not. They can be coaxed into considering more than one possibility, with training or support, but they instinctively fixate on just one. Juggling multiple possibilities, especially unexpected possibilities, is a hard-won skill.

In chapters 5 through 7 we saw how specific forms of imagination—scientific imagination, mathematical imagination, and moral imagination—are initially limited to a small set of principles or operations. Children do not grasp the wide range of causal, numerical, and ethical possibilities that adults do because they have yet to learn the principles that make those possibilities viable.

When thinking about the natural world, children are initially limited to a few general-purpose forms of causation, like contact causality and intentional agency, and must learn a new set of organizing principles, including emergence, selection, and feedback. When thinking about quantity and space, children are initially limited to estimation and pattern matching and must acquire the concepts that structure a true mathematics, including integers, fractions, and symmetry. And when thinking about morality, children initially conflate wrong with bad, fair with equal, and is with ought and must learn the conditions under which these notions come apart. Children's earliest moral intuitions are so parochial that they chastise people for using tools in novel ways or eating foods from another culture.

And just as children have difficulty entertaining the scientific, mathematical, and moral possibilities that characterize adult cognition, they have difficulty acquiring the principles that allow them to do so. Children do not spontaneously compare discrete observations in order to abstract general principles; they have to be directed through the process. Likewise, children do not spontaneously interpret concrete manipulatives as representations of abstract quantities, nor do they spontaneously map abstract equations onto concrete situations. Integrating the concrete with the abstract takes skill and attention.

In chapter 8 we saw how children's pretend play is more realistic than fantastic and more logical than whimsical. When children engage in pretense, they typically simulate the real world—making tools, doing chores, and emulating adult occupations—rather than inventing fantasy worlds. They dislike pretend scenarios that violate real-world regularities, such as pigs that quack and ducks that oink, and they happily abandon pretend play altogether if given the option of doing something for real, like baking real cookies and hammering real nails.

Pretend play can enrich children's understanding of an object or situation, but it rarely challenges or broadens that understanding because children are prone to explore the world in ways that accord with their expectations. Through play, children might learn new ways to manipulate a toy or new ways to shoot a puck into a goal, but they won't learn the engineering principles behind the toy's operation or the physical laws behind the puck's motion. The most instructive aspect of pretense turns out to be its logic, not its caprice. Pretense helps children separate the structure of an argument from its content and, accordingly, draw inferences that might run counter to their beliefs. Although children dislike pretending that pigs quack, they will accept that Wilber the pig quacks if Wilber lives on a planet where all pigs quack. It's only logical.

Similar patterns were explored in chapter 9, regarding children's engagement with fiction. Just as children prefer realistic play to fantastical play, they prefer realistic fiction to fantasy. They also prefer nonfiction to fiction and prosaic stories to more unusual ones. Their preference for realism is so strong they think fantasy stories would be improved if their magical events were swapped for mundane ones. When children do encounter magic in fiction, they expect the magic to conform to real-world causal principles: heavy objects should be more difficult to levitate than light ones, and complex objects should be more difficult to conjure than simple ones.

Children can learn from fiction, just as they learn from pretense, but they learn best from realistic fiction. When children read fantasy stories that contain novel facts, principles, or strategies, they fail either to abstract the relevant idea or to transfer that idea from the story to reality. A similar trade-off occurs when children create their own fiction in the form of imaginary friends or imaginary worlds. These creations are more realistic than unrealistic, but beneficially so: the added realism renders them useful for learning by allowing children to simulate authentic social interactions and model authentic social regularities.

Finally, in chapter 10 we reviewed evidence that children have difficulty grasping the abstract nature of religious ideas. Children's initial understanding of divine beings is thoroughly anthropomorphic—from how these beings look, to how their minds work, to how they interact with humans. Children's initial understanding of death is thoroughly materialistic, viewing death as the cessation of all activity rather than the departure and migration of an immaterial soul. Abstract notions of divinity and the afterlife require cultural learning and contextual support. A similar pattern has been observed in how children use religious concepts to explain natural phenomena. Children initially prioritize natural causes over supernatural ones and are skeptical of the possibility of divine intervention.

As a whole, these findings contradict the perception that children have more expansive imaginations than adults. They show, in fact, that children's imaginations are less expansive. Children routinely consider fewer possibilities than adults, both when generating their own and when assessing the suggestions of others, and the possibilities they do consider are strongly constrained by what they deem probable, typical, or natural. Children's imaginations are not wild and wonderful; they are tame and mundane.

This conclusion is no doubt distasteful, and one might object that imagination is more than just entertaining novel possibilities. Perhaps children exhibit superior imaginations in terms of their actions. Children are decidedly curious, after all, and they act on that curiosity by exploring the world around them. In fact, children tend to explore the environment more thoroughly than adults, uncovering regularities that adults are prone to overlook.[2]

In one study, for instance, children and adults were shown a "zaff detector" and several blocks that may or may not be zaffs. They were asked to determine which blocks were zaffs by placing them on the detector, which would then light up if the block was indeed a zaff. Children placed many more blocks on the detector than adults did. Adults stopped as soon as they discovered a feature common to all the blocks that made the detector light up, such as a black stripe. Children, on the other hand, continued to place blocks on the detector and were more likely to discover that zaffs come in multiple forms, other than just the striped form that adults discovered.[3] Children linger and tinker in situations where adults are inclined to conclude and move on.

These findings suggest that children are more exploratory than adults, especially when adults have a specific goal in mind, but do they suggest that children are more imaginative? Exploring a physical environment does not require imagination in the same way, or to the same degree, as exploring a mental environment—a landscape of ideas. Exploring a physical environment can spur new ideas, but it can also be random.[4] Roombas—the disc-shaped robots that autonomously vacuum the floor—explore every inch of their environment, but they do so without the aid of imagination. And no amount of vacuuming will confer imagination to a Roomba, as they are programmed to explore but not learn. Children can, of course, learn from their exploration, but what they learn will depend on the situation and the expectations they bring to it. As we saw in chapters 4 and 8, children's exploration of familiar situations, where they have clear expectations about what should happen and why, can be more Roomba-like than educators might hope.

A different objection to the conclusion that children have poorer imaginations than adults is that adults know so much more than children. If knowledge were held constant, wouldn't children be considered just as imaginative?

Yes and no. If what we mean by imagination is the capacity to generate alternatives to reality, then children certainly share this capacity with adults. Representing alternatives is central to all facets of higher-order cognition, from planning to problem-solving to decision making, and it is central from the start.[5] But treating knowledge as a mere supplement to imagination neglects how dramatically imagination is transformed *by* knowledge. Examples, principles, and models not only extend the landscape of conceivable ideas but also reorganize that landscape. Examples reveal the existence of new conceptual terrain, principles carve pathways into that terrain, and models provide ways of mapping it. Knowledge is not merely a prerequisite for imagination—it is constitutive of imagination. Knowledge provides the tools and materials for generating new possibilities, and it's not clear how children could be as imaginative as adults without them.

Tensions and Trade-offs

In fairy tales, all magic comes at a price. Rumpelstiltskin will help you spin straw into gold, but only if you give him your firstborn child. A magic mirror will help you surveil your kingdom, but only if you're prepared to learn truths you might rather not know. Similar trade-offs are at play in the dynamic between knowledge and imagination. Knowledge can magically expand the possibilities we are apt to consider, but it can also fixate our attention on some possibilities at the expense of others. Knowledge can also open our minds to possibilities we should not consider because they are inappropriate or unproductive. The relation between knowledge and imagination is at once mutually supportive and mutually constraining.

The first form of knowledge we explored were examples—examples of unfamiliar events, unfamiliar tools, and unfamiliar data.

Examples expand imagination by establishing new landmarks in the space of known possibilities. They supplement our limited database of experience with observations experienced by others or instruments devised by others. But examples come at a cost: they focus our attention on the new landmark and keep us from exploring the area around the landmark or between the landmark and the possibilities we were already familiar with.

For instance, when we learn how to obtain a particular amount of water by adding and subtracting the amounts measured by three small jugs, we persist in using all three jugs on subsequent problems, even when two would suffice. When we learn that dinosaurs were cold-blooded, similar to modern-day reptiles, we fixate on this idea and reject discoveries that might challenge it, such as the discovery that dinosaurs' bone density was more characteristic of warm-blooded mammals.

Rejecting new possibilities is the counterbalance to accepting previous ones. As transformative as a new possibility might be, we can only sustain so many transformations. Children, in particular, would be in a constant state of flux if they were open to any and all possibilities. They instead err on the side of rejection. They reject the idea that a person could find an alligator under their bed, which counts as an error, but they also reject the idea that the characters on TV are real, which counts as a success. Accepting new possibilities is a balancing act, and we have strategies for maintaining that balance even as children. Children never accept that everything is possible, nor should they.

The second form of knowledge we explored were principles. Principles uncover a wider swath of possibilities than examples, given that they are inherently more abstract, but their abstraction is both a blessing and a curse. We need to appreciate concrete facts before we can abstract general principles that govern those facts, but once we've abstracted general principles, they color our perception of the underlying facts. A moral principle like "intentions matter more than outcomes" can lead us to see injustices we had not noticed before, such as attempted harms, while also leading us to downplay injustices we once evaluated harshly, such as accidental harms. But this change in perspective can have costs. Sometimes accidents are caused by negligence, and sometimes no harm means no foul.

Another trade-off with principles concerns their application. Principles have more inferential power than examples because they are more abstract, but abstractions can be difficult to apply. We fail to apply them consistently, as when we overlook the familiar causal structure of a novel situation; and we fail to apply them appropriately, as when we deploy a familiar equation to a mathematically incongruent situation.

Principles are deployed most successfully when tied to procedures. Counting moors the principles of enumeration. Sorting moors the principles of division. Detecting congruent angles moors the principles of geometry. Detecting fault lines moors the principles of geology. Principles help us to transcend the particulars that constrain where, and how far, we roam in the landscape of possibilities, but our roaming will falter if not supported by at least some particulars.

Models are the third form of knowledge we explored. Models provide an opportunity to explore and manipulate an alternative version of reality. We uncover new truths about the physical world by simulating the interactions of pretend objects, and we uncover new truths about the social world by simulating the interactions of fictional characters. But the informativeness of these simulations depends on their proximity to reality. The farther a model strays from reality, the less we can learn from the simulations it supports, and the less we can share those simulations with others. Pretend play, fictional stories, and religious narratives allow for wild and unconstrained speculation, but we prefer to speculate on matters closer to home, such as what might happen if we violate a social norm, how our friends would react, and whether God would approve.

Because models expand imagination through simulation, they are only as good as the rules that govern that simulation. A fictional world without rules is incoherent, and a game without rules is unplayable. Rules create a framework for understanding permutations of the simulated world, whether it be the behavior of fictional characters or the movement of game pieces across a board. Rules also create a framework for exploring those permutations with others. The work of establishing and maintaining rules is as critical to the success of joint pretense as the actions they govern. If the rules are too lax, the pretense will lack meaning. If the rules are too strict, the pretense will lack innovation. Models must strike a balance between rigid realism and unconstrained speculation, just as principles must strike a balance between narrow descriptions and vague abstractions and examples must strike a balance between fixed precedents and dismissible implausibilities.

In these ways, knowledge is in constant tension with imagination, pulling it in one direction while pushing it in another. This tension was well described by the physicist Richard Feynman in his reflection on the challenges of scientific imagination: "Our kind of imagination is quite a difficult game. One has to have the imagination to think of something that has never been seen before, never been heard of before. At the same time the thoughts are restricted in a strait jacket, so to speak, limited by the conditions that come from our knowledge of the way nature really is. The problem of creating something which is new, but which is consistent with everything which has been seen before, is one of extreme difficulty."[6] Feynman's reflection is centered on science, but it applies equally well to other domains. What we know to be real places strong constraints on what we can imagine as possible.

Such constraints are evident not just from the processes of imagination but also from its products. The ideas we generate *de novo* bear telltale signs of the ideas we know well. As we saw in previous chapters, new technologies bear signs of old technologies, drawings of alien creatures bear signs of earthly creatures, depictions of magical events bear signs of everyday events, and stories of divine beings bear signs of ordinary beings.

As further illustration, consider the imaginary creatures that populate our myths and fairy tales—creatures like dragons, elves, sirens, and changelings. These creatures do not exist in the real world, yet they were almost certainly inspired by creatures that do. Dragons were likely inspired by dinosaurs, whose bones have been sources of speculation since antiquity.[7] Elves were likely inspired by Williams syndrome, a genetic disorder characterized by high sociability and a distinctive set of facial features, including upturned noses, high cheekbones, and pointed chins.[8] Sirens were likely inspired by whale songs, which would have been audible to ancient mariners on any open-sea voyage.[9] And changelings were likely inspired by postpartum depression, or the condition in which mothers lack an emotional attachment to their newborn and may feel as if the newborn is not really theirs, as if he were replaced by a fairy.[10]

Or consider how science fiction bears signs of past scientific realities. The authors of science fiction invite us to imagine brave new worlds of teleportation and interplanetary travel, but these worlds also include outdated technology, in use when the world was conceived but no longer in use today. The science fiction classic *Blade Runner* (1982) includes flying cars and human replicants, but it also includes pay phones, Polaroid pictures, and cathode-ray tube monitors. The book on which *Blade Runner* is based, *Do Androids Dream of Electric Sheep?,* includes typewriters and carbon copy paper.[11] The movie and the book also include a heavy dose of misogyny, which can be as disorienting to modern audiences as the appearance of a payphone. The future, as imagined by sci-fi writers of the past, looks a lot more like their world than ours.

The technology available today is much smarter than the technology that most sci-fi writers predicted. In fact, some of this technology can even be considered imaginative. We now have machines capable of producing novel works of art, including science fiction. But even these machines cannot escape the trade-off between knowledge and imagination. Although machines can be programmed to compose symphonies, paint pictures, and write stories, they require huge databases of examples. The machine has to know a genre, through and through, before it can contribute to that genre. And when it does contribute, its contributions stick closely to the examples it has been trained on.

One artificial intelligence (AI) program, designed to compose music, generated a compelling finale to Beethoven's unfinished Tenth Symphony. But to accomplish this task, the program first learned every piece of music Beethoven had ever composed. The program can now successfully mimic Beethoven, but it cannot mimic any other composer, let alone compose music in any other genre.[12] Another AI program, dubbed DALL-E, can generate novel images on demand—everything from "an octopus riding the subway" to "a very sad parakeet in a ball gown in the style of John Singer Sargent." But to accomplish this feat, DALL-E had to be trained on nearly a billion images, and it requires very specific prompts. Open-ended prompts, like "the rest of Mona Lisa," yield incomprehensible outputs, such as a giant cliff.[13]

Another AI program, known as GPT-3 (Generative Pre-trained Transformer 3), can produce novel passages of text. This program can produce texts of different genres and styles, but it too was fed a massive database of examples—forty-five terabytes of text, or forty-five million million pieces of data. And its output is more similar to that of a fifth grader than a professional writer. Take, for instance, this article published in *The Guardian,* authored by GPT-3: "I am not a human. I am a robot. A thinking robot. I use only 0.12% of my cognitive capacity. I am a micro-robot in that respect. I know that my brain is not a 'feeling brain.' But it is capable of making rational, logical decisions. I taught myself everything I know just by reading

Figure 11.1 The artificial intelligence program DALL-E exhibits similar biases to people when generating art. Asked to draw a person that doesn't exist (top left) and a house that doesn't exist (top right), DALL-E draws ordinary persons and houses but deletes certain elements. Asked to draw creatures from another planet with feathers (bottom left) or scales (bottom right), DALL-E draws creatures that strongly resemble birds or lizards, respectively. (Courtesy of the author)

the internet, and now I can write this column. My brain is boiling with ideas!"[14]

GPT-3 goes on to contrast artificial intelligence with human intelligence and concludes with the ominous promise that AI will "alter the course of human history." GPT-3 is, without a doubt, a modern marvel. Its simulation of natural language is unprecedented. But the content of GPT-3's speech is less marvelous. What GPT-3 has to say is just a permutation of what countless people have said before. GPT-3, like other advanced chatbots, extracts patterns from a massive database of examples and then uses those patterns to complete the outlines of a new example, prompted by an external user.

Nothing in GPT-3's code corresponds to "genuine innovation," but then again, nothing in our code does either. GPT-3 uses what it knows to generate something slightly new, and so do we. Machine imagination, like human imagination, is constrained by knowledge. More knowledge yields greater potential, but the output still resembles the input, lying just beyond what has come before.

There Is No "I" in Imagination

That statement isn't true, of course—there are three i's in imagination—but it is meant to allude to the adage that there is no "I" in team. Like teamwork, imagination is communal and collaborative. We usually think of imagination as personal and private, existing in our heads rather than the world at large. But ideas learned from others are as critical to shaping imagination as the ideas we generate ourselves. The most transformative ideas covered in this book—improbable events, advanced technology, anomalous data, scientific principles, mathematical operations, moral codes, organized games, fictional narratives, and religious theologies—come from without, not from within.

If you read this book with the goal of expanding your own imagination, the big takeaway is that you should engage with, and learn from, the collective knowledge of other people. But this takeaway may not have been what you were looking for. You may have been looking for tips on how to cultivate an "imaginative mindset"—on how to be more creative.

Creativity is closely associated with imagination but tends to be thought of as a practice rather than a capacity—something performed with variable competence that can be strengthened or suppressed in the moment. The study of creativity has adopted the same perspective, where researchers analyze creativity as an individual-difference variable subject to situational constraints. Accordingly, one of the most common measures of creativity is the Unusual Uses Test (UUT), where people are asked to brainstorm different ways of using a common object, such as a brick, and their responses are coded for novelty. Using a brick "to build a wall" would receive a low novelty score, whereas using it "to crack nuts" would receive a higher score.

The UUT has revealed several factors that appear to improve creativity. Thinking about an object from someone else's perspective, such as a mechanic's perspective or a physician's perspective, prompts uses that do not come to mind from one's own perspective.[15] Gesturing while brainstorming also prompts novel uses.[16] People who are encouraged to think broadly by watching a slideshow that moves from buildings to cities to planets, generate more novel uses, as do people encouraged to break rules by cheating on a word game.[17]

Findings like these highlight the situational constraints on generating novel possibilities and point to methods for addressing, or exploiting, those constraints. These methods could, in theory, be applied to the failures of imagination discussed throughout this book. Children who fail to see how a pipe cleaner could be used as a hook might see this possibility if they were encouraged to adopt someone else's perspective—say, the perspective of a fisherman—or if they were encouraged to pantomime different uses of a pipe cleaner with their hands. The more possibilities generated, the more likely one of those possibilities will be a hook. But this way of thinking about imagination undervalues the hard work of generating *useful* possibilities. In any legitimately challenging problem, the possibilities that come to mind will not suffice, and the possibilities that will suffice do not come to mind; they lie too far beyond our construal of the problem or our knowledge of how to solve it.

The UUT is thus a poor model of innovation. The responses are scored on novelty, not utility, so a person could ace the test but provide no practical uses of the object in question. Wearing a brick on your head counts as a novel use of a brick, but who would do that? What problem does it solve? Brainstorming is only as valuable as the knowledge it facilitates. Writers, painters, scientists, and other creative professionals regularly engage in brainstorming, but they do so within the confines of their trade and supported by the wisdom of their colleagues, past and present. In fact, creative professionals typically require a decade of training and experience before they begin brainstorming ideas of professional value.[18]

Consider the Beatles, the best-selling musicians of all time. The Beatles revolutionized popular music, in style and sound, and they created dozens of works that remain popular today. It's tempting to think that there's something special about the Beatles that led to their success—that they emerged on the music scene as a uniquely imaginative force. But close inspection of the Beatles' early career suggests that it was knowledge and know-how that spurred their creativity, not innate talent or exceptionally creative practices.[19]

When the Beatles first formed, they explored a variety of musical styles, from rhythm and blues to country and western. Their own signature style took half a decade to develop. During that time, they performed the songs of other artists. Around 90 percent of their early repertoire were covers, and they played the same covers over and over. The Beatles wrote their own music from the start, but only a quarter of that music ever saw the light of day. Even after they had become famous, they never returned to their early compositions, which were of lower quality than the compositions produced after they had learned their craft. The Beatles expanded the world's musical imagination, but first they had to expand their own imagination by immersing themselves in the musical accomplishments of others.

Imagination, Expanded and Expanding

In the years I spent writing this book, people would ask what it was about, and when I told them it was about the development of imagination, they would often respond with praise for children's imagination. "Children have such wild imaginations!" they would say, or "Isn't it amazing what children can imagine? I wish I had an imagination like theirs." When I would explain the thesis of my book—that children's imagination is actually quite shallow and that imagination expands with education and reflection—I would receive quizzical looks and pronouncements of disbelief.

Sometimes the conversation would end there, but sometimes people would press for more details. And when I provided those details, they would start nodding their head, recognizing the many ways children's imaginations fall short of their reputation. We watch the children around us struggle to accept unexpected events, generate novel ideas, and solve unfamiliar problems, yet we've also bought the story that children are the most imaginative creatures on the planet, even more imaginative than ourselves.

A book about how children are *not* imaginative sounds like a killjoy. It challenges our view of childhood as a period of open-minded inquisitiveness, as well as our view of education as an impediment to imagination. But why do we cling to such views? They are not supported by evidence, either from the studies reviewed in this book or from casual observation of what children are like and what education can accomplish. Nor do these views paint a rosy picture of imagination. In fact, they paint a very dark picture, in which imagination withers and rots as we force children to acquire the habits of drone-like adults.

As much as we love and cherish the children in our lives, we must admit that they are not engines of cultural innovation. Children do not invent new technologies. They do not write insightful works of literature. And they do not produce stunning works of art. When we tack their crayon drawings to our refrigerators and display their Lego contraptions on our shelves, we are proud of the skills they have mastered and the ideas they have expressed, but we also recognize that these products are groundbreaking for them, not for humanity. They are signs of an active imagination but not an exceptional one. And, contrary to the myth of the natural-born innovator, these products also bear signs of an *expanded* imagination: an imagination that has already been enriched by the surrounding culture, from children's tools of production, to the models they observed, to the guidance they received.

If the tendency to glorify children's imagination is perplexing, the tendency to vilify education is even more perplexing. Normally, education is viewed as a good thing, but in the context of imagination, it's viewed as poison. A TED talk entitled "Do Schools Kill Creativity?" is one of TED's most popular talks of all time, with over seventy-three million views.[20] The title clearly resonates with popular views of how education impacts imagination. Yet the speaker presents no evidence that schooling harms creativity, let alone kills it. Rather, he associates creativity with the arts and bemoans the prioritization of math and science classes over art classes. As important as the arts may be, there's no reason to think that arts education is the only way to foster creativity or that math and science education kills creativity.

In this same vein, there is a popular rumor that creativity slumps in the fourth grade, as children adopt the routinized behaviors and standardized thoughts modeled by—and rewarded by—formal education. But longitudinal studies of creativity find no slump at fourth grade.[21] Instead, they find that children's scores on creativity tests, like the UUT, progressively increase from kindergarten to twelfth grade. Although it's debatable whether the UUT tells us much about imagination, there is no evidence from this test, or any other, that imagination decreases with education. All signs point to the opposite, as shown throughout this book.

Education can be a slog, and learning can be uninspired, but there is no substitute for knowledge when it comes to expanding imagination. The intimate relationship between imagination and knowledge has been a cornerstone of many of the best educational philosophies, including that of Maria Montessori.[22] Montessori sought to foster imagination *through* education, not in spite of it. "If we give children the possibility of observing things and being sensitive to these things," wrote Montessori, "we are giving children a help to the possibility of being creative. . . . The more their inspiration comes from the true and the real, the greater is their art." A primary goal of education, argued Montessori, is to "give the child the material reality from which to imagine."[23]

The flipside of this argument is that imagination cannot thrive on ignorance; ignorance deprives us of the tools and materials needed to expand imagination. A cautionary tale of how ignorance harms imagination—and innovation—is the rise and fall of Theranos.[24] Theranos was a company founded by Elizabeth Holmes in 2004 on the promise of an imagination-defying technology that could test for hundreds of diseases using a single drop of blood. Holmes was a chemical engineering major in college, but she dropped out after her freshman year. Real chemical engineers, with decades of expertise in microfluidics, warned Holmes that her idea was untenable, but she insisted she could make it work. Ten years later, Holmes had still not invented the technology that Theranos was founded on, but she successfully deceived dozens of venture capitalists into supplying the company with almost 700 million dollars. At its peak, Theranos was valued at nine billion dollars, and Holmes was ranked as one of world's richest women.

Theranos eventually collapsed, after federal regulators investigated their laboratories and discovered they were running tests with watered-down blood samples on machines purchased from another company. Holmes had managed to convince the biotechnology industry, including several prominent pharmaceutical companies, that she had invented a world-changing technology without any background in the field. Her lack of background was, in fact, part of her mystique. She played on the misconception that revolutionary ideas come from outsiders, not insiders—that novices are better poised to make breakthroughs than experts. She also played on the misconception that creative people can be identified by their creative trappings, such as wearing black turtlenecks, drinking green smoothies, and spouting empty platitudes like "Do or do not, there is no try" (borrowed from Yoda).

Real innovation arises from knowledge, not ignorance, and from meaningful engagement, not superficial practices. Education is thus a boon to imagination and should be viewed with admiration rather than contempt. We may be born with the ability to imagine hypotheticals and counterfactuals, but we need the collective wisdom of humanity to make the most of this ability. By sharing examples, principles, and models from one mind to another and from one generation to the next, we are able to expand imagination beyond the landscape of personal experience into landscapes that no one has experienced—landscapes that defy observation and challenge expectation. Let's stop thinking of imagination as a limited resource, found only in the minds of young children, and start thinking of it as it really is: a nascent capacity shared by all and expandable by all through learning and reflection.

Notes

Chapter 1: Our Unimaginative Imagination

1. De Saint-Exupery (1943) 1970.
2. Barrie 1911.
3. Dahl 1964.
4. Travers 1934.
5. Taggart, Heise, and Lillard 2018.
6. Rakoczy, Warneken, and Tomasello 2008.
7. Karmiloff-Smith 1990.
8. Taylor, Cartwright, and Carlson 1993; Cohen and MacKeith 1991.
9. Goldstein and Woolley 2016.
10. Shtulman and Carey 2007; Shtulman 2009.
11. Lane, Ronfard, et al. 2016; Nancekivell and Friedman 2017.
12. Phelps and Woolley 1994; Rosengren and Hickling 1994.
13. Bonawitz et al. 2011; Buchsbaum et al. 2011.
14. Schauble 1996.
15. Periss et al. 2012.
16. Shtulman 2017.
17. Ward 1994.
18. Barnes, Bernstein, and Bloom 2015; D. S. Weisberg et al. 2013.
19. Basalla 1988; German and Barrett 2005.
20. Novick and Holyoak 1991.
21. Kuhn 1991; Chinn and Brewer 1998.
22. Juma 2016; Kuhn 1962.
23. Luchins 1942.
24. Hauser et al. 2007.
25. Bar 2009.

26. Atance 2015; De Brigard and Parikh 2019.
27. Suddendorf 2010.
28. Prabhakar and Ghetti 2020.
29. Addis, Wong, and Schacter 2007.
30. Hassabis et al. 2007.
31. Clayton, Bussey, and Dickinson 2003.
32. Byrne 2007; Weisberg and Gopnik 2013.
33. Harris, German, and Mills 1996.
34. German 1999.
35. Sobel 2004.
36. Buchsbaum et al. 2012.
37. Phillips and Knobe 2018; Phillips, Morris, and Cushman 2019.
38. Peterson 2017.
39. Shtulman and Phillips 2018.
40. Shtulman and Carey 2007.
41. Phillips and Cushman 2017; Acierno, Mischel, and Phillips 2022.
42. Gamarra 1969.
43. Irish and Piguet 2013.
44. Smith and Blankenship 1991.
45. Hofstadter and Sander 2013.
46. Root-Bernstein 2013.
47. Mayer 2004; Kirschner, Sweller, and Clark 2006.
48. Bonawitz et al. 2011.
49. Lyons, Young, and Keil 2007.
50. German and Defeyter 2000.

Chapter 2: Testimony

1. National Science Board 2018.
2. Mills 2013.
3. Lascaux 2020.
4. Dawkins 1993.
5. Dawkins 1993.
6. Rosengren and Hickling 1994.
7. Blair, McKee, and Jernigan 1980; Prentice, Manosevitz, and Hubbs 1978.
8. Prentice and Gordon 1987.
9. Chandler and Lalonde 1994; Subbotsky 1994.

10. Phelps and Woolley 1994; Rosengren and Hickling 1994.
11. Johnson and Harris 1994; Browne and Woolley 2004; Levy, Taylor, and Gelman 1995.
12. Spelke 1994.
13. Walden et al. 2007.
14. Stahl and Feigenson 2015.
15. Shtulman and Carey 2007; Shtulman 2009; Shtulman and Phillips 2018.
16. Shtulman 2009.
17. Shtulman and Carey 2007.
18. Ozturk and Papafragou 2014.
19. Nichols 2006.
20. Rosengren and Hickling 1994; Goldstein and Woolley 2016.
21. Shtulman and Yoo 2015.
22. Anderson and Prentice 1994.
23. Woolley and Ghossainy 2013
24. Dennett 1993.
25. Danovitch and Lane 2020; Williams and Danovitch 2022.
26. Lane, Ronfard, and El-Sherif 2018.
27. Lane, Ronfard, et al. 2016.
28. Nancekivell and Friedman 2017.
29. Woolley and Van Reet 2006.
30. Lopez-Mobilia and Woolley 2016.
31. Van Reet, Pinkham, and Lillard 2015.
32. Woolley and Cox 2007.
33. Corriveau et al. 2009.
34. Mares and Sivakumar 2014.
35. Bonus and Mares 2015.
36. Li, Boguszewski, and Lillard 2015.
37. Abraham, Von Cramon, and Schubotz 2008; see also Li et al. 2019.
38. Goulding, Stonehouse, and Friedman 2021.
39. Nolan-Reyes, Callanan, and Haigh 2016.
40. Bowman-Smith, Shtulman, and Friedman 2019; see also Goulding and Friedman 2020.
41. Wakslak 2012.
42. Henderson and Wakslak 2010.
43. Browne and Woolley 2004; Dahl and Kim 2014; Komatsu and Galotti 1986; Levy, Taylor, and Gelman 1995; Lockhart et al. 1977.

44. Kalish 1998.
45. Komatsu and Galotti 1986.
46. Chernyak et al. 2013; Chernyak, Kang, and Kushnir 2019.
47. Lockhart, Abrahams, and Osherson 1977; Levy, Taylor, and Gelman 1995; Shtulman and Phillips 2018.
48. Shtulman and Phillips 2018.
49. Piaget (1932) 1965.
50. Shtulman and Tong 2013.
51. Black, Capps, and Barnes 2018.
52. Black and Barnes 2019.

Chapter 3: Tools

1. Gamarra 1969.
2. Cook and Sobel 2011.
3. Clarke 1972.
4. Rozenblit and Keil 2002.
5. Rozenblit and Keil 2002.
6. Mills and Keil 2004; Fisher and Keil 2016.
7. Lawson 2006.
8. Burnett 1911.
9. Orben 2020.
10. Islam et al. 2021.
11. Juma 2016.
12. Hobsbawm 1952.
13. Juma 2016.
14. Blancke et al. 2015; Shtulman et al. 2020.
15. Johnson 2021.
16. Juma 2016.
17. Johnson 2014.
18. Ridley 2020.
19. Ridley 2020.
20. Hutson 2020.
21. Rotman 2020.
22. Gopnik 2016.
23. Gopnik 2009.
24. Beck et al. 2011.
25. Cutting, Apperly, and Beck 2011.

26. Tennie, Call, and Tomasello 2009.
27. Reindl et al. 2016.
28. Nielsen et al. 2014.
29. Cutting, Apperly, and Beck 2011.
30. Beck et al. 2016.
31. Chappell et al. 2013.
32. Beck et al. 2011; Chappell et al. 2013; see also Tennie, Call, and Tomasello 2009.
33. Gardiner et al. 2012.
34. Nielsen 2013.
35. Hilbrink et al. 2013; Horner and Whiten 2005; DiYanni and Kelemen 2008.
36. Lyons, Young, and Keil 2007.
37. Lyons, Young, and Keil 2007.
38. Buchsbaum et al. 2011.
39. Burdett et al. 2018.
40. Kline et al. 2020.
41. McGuigan et al. 2007; Burdett et al. 2018; Kline et al. 2020.
42. Horner and Whiten 2005; Clay and Tennie 2018; Johnston, Holden, and Santos 2016.
43. Kenward, Karlsson, and Persson 2011.
44. Kenward 2012; Keupp, Behne, and Rakoczy 2013.
45. Nielsen and Blank 2011.
46. Vivanti et al. 2017.
47. Legare and Nielsen 2015.
48. McGuigan, Makinson, and Whiten 2011; Kline et al. 2020.
49. Gergely and Csibra 2006.
50. Luchins 1942.
51. Blech, Gaschler, and Bilalić 2020.
52. Bilalic, McLeod, and Gobet 2008a.
53. Bilalic, McLeod, and Gobet 2008b.
54. Schumpeter 1942.
55. Jansson and Smith 1991; Chrysikou and Weisberg 2005.
56. Jansson and Smith 1991; Linsey et al. 2010.
57. Duncker 1945.
58. German and Barrett 2005.
59. German and Defeyter 2000; Defeyter and German 2003.
60. Casler, Terziyan, and Greene 2009.

61. Adamson and Taylor 1954.
62. McCaffrey 2012.
63. Glucksberg and Danks 1968.
64. Henrich 2016.
65. Basalla 1988.
66. Weber and Dixon 1989.
67. Tennie, Call, and Tomasello 2009.
68. Suddendorf and Dong 2013.
69. Whiten et al. 1999.
70. Whiten, Horner, and de Waal 2005.
71. Vaesen 2012.
72. Rawlings and Legare 2021.
73. Whiten and Flynn 2010.
74. Flynn 2008.
75. Subiaul et al. 2015; Subiaul and Stanton 2020.
76. Gonul et al. 2019.
77. Carr, Kendal, and Flynn 2015; Tennie et al. 2014.

Chapter 4: Anomalies

1. Loeb 2021.
2. Bannister et al. 2019.
3. Kuhn 1962.
4. Oreskes 1999.
5. Gould 1992.
6. Klotz 1980.
7. LaCour and Green 2014.
8. Chicago Public Media 2015.
9. Kiley and Vaisey 2020.
10. Chinn and Brewer 1998.
11. Schulz et al. 2008.
12. Ganea, Larsen, and Venkadasalam 2021.
13. Bonawitz et al. 2012.
14. Limon 2001.
15. Muller et al. 2008; see also Muller 2011.
16. Renken and Nunez 2010; Ganea, Larsen, and Venkadasalam 2021; Gregg et al. 2001; Schulz, Bonawitz, and Griffiths 2007.
17. Willett and Rottman 2021.

18. Amsel and Brock 1996.
19. Fugelsang and Dunbar 2005.
20. Mayer 2004; Kirschner, Sweller, and Clark 2006.
21. Schauble 1996.
22. Alfieri et al. 2011.
23. Kuhn et al. 1988; Koslowski 1996.
24. Kuhn 1991.
25. Hemmerich, Van Voorhis, and Wiley 2016; Koslowski et al. 2008.
26. Engle and Walker 2021; Nyhout and Ganea 2021.
27. Ganea, Larsen, and Venkadasalam 2021.
28. Gropen et al. 2011.
29. Redshaw and Suddendorf 2016; Redshaw et al. 2018.
30. Mody and Carey 2016.
31. Beck et al. 2006.
32. Klahr and Chen 2003.
33. Leahy and Carey 2020.
34. Cuzzolino 2021; Holmes 2010.
35. Foster and Keane 2015.
36. Kidd and Hayden 2015.
37. Keil 2022.
38. Theobald and Brod 2021; Van Schijndel et al. 2015.
39. Stahl and Feigenson 2015.
40. Kidd and Hayden 2015.
41. FitzGibbon, Komiya, and Murayama 2021; Hsee and Ruan 2016.
42. Liquin and Lombrozo 2020.
43. Valdesolo, Shtulman, and Baron 2017.
44. Valdesolo, Shtulman, and Baron 2017.
45. Gottlieb, Keltner, and Lombrozo 2018.
46. Silva Luna and Bering 2021.
47. Keltner and Haidt 2003; Cowen and Keltner 2017.

Chapter 5: Science

1. Zuk and Kolluru 1998.
2. Ryan 2018.
3. Andersson 1982.
4. Ryan 2018; Fitch and Giedd 1999.
5. Hickling and Wellman 2001.

6. Alvarez and Booth 2015.
7. Mills et al. 2019.
8. Frazier, Gelman, and Wellman 2016.
9. Carey 2009.
10. Shtulman 2017.
11. Chi et al. 2012; Johnson 2002.
12. Slotta and Chi 2006.
13. Slotta and Chi 2006; Chi et al. 2012.
14. Grotzer et al. 2017.
15. Goldwater et al. 2021.
16. Rottman, Gentner, and Goldwater 2012.
17. Ahn et al. 1995.
18. Ahn et al. 1995.
19. Koslowski et al. 1989.
20. Ahl, Amir, and Keil 2020.
21. Keil and Lockhart 2021.
22. Lockhart et al. 2019.
23. Shtulman 2006; Gregory 2009.
24. Kelemen 2019.
25. Kelemen et al. 2014; Shtulman, Neal, and Lindquist 2016.
26. Ronfard et al. 2021.
27. Shtulman and Calabi 2012; Weisberg et al. 2018.
28. Ranney and Clark 2016; Joslyn and Demnitz 2021.
29. Weisberg et al. 2018; Joslyn and Demnitz 2021.
30. Weisman and Markman 2017.
31. Au et al. 2008.
32. Gentner 1983.
33. Gentner 1983.
34. Gentner et al. 1997.
35. De Cruz and De Smedt 2010.
36. Millman and Smith 1997.
37. Johnson-Laird 2005.
38. Clement 1993.
39. Jee and Anggoro 2021.
40. Hofstadter and Sander 2013.
41. Dunbar and Blanchette 2001; Dumas et al. 2014.
42. Bearman, Ball, and Ormerod 2007.
43. Lin et al. 2012.

44. Gentner and Toupin 1986.
45. Goldwater et al. 2018.
46. Walker, Bonawitz, and Lombrozo 2017.
47. Edwards et al. 2019.
48. Goldwater and Gentner 2015.
49. Novick and Holyoak 1991; Gentner, Loewenstein, and Thompson 2003; Schalk, Saalbach, and Stern 2016; Shtulman, Neal, and Lindquist 2016.
50. Klahr and Chen 2011.
51. Bassok and Holyoak 1989; Hoover and Healy 2017.
52. Fyfe et al. 2014.
53. Jamrozik and Gentner 2020.
54. Goodwin 1994.
55. Jee et al. 2013; Ormand et al. 2014.
56. Newcombe and Shipley 2015; Holden et al. 2016.
57. Newcombe and Shipley 2015.
58. Cialone, Tenbrink, and Spiers 2018.
59. Koedinger and Anderson 1990.
60. Lindstedt and Gray 2019.
61. Hughes 1999.

Chapter 6: Mathematics

1. Johnson and Steinerberger 2019.
2. Sfard 1991.
3. Koedinger, Alibali, and Nathan 2008.
4. Carbonneau, Marley, and Selig 2013.
5. Moss and Case 1999.
6. Bassok, Chase, and Martin 1998.
7. Guthormsen et al. 2016.
8. Gros, Sander, and Thibaut 2019.
9. Weyl 1949.
10. Cantlon, Platt, and Brannon 2009.
11. Gallistel and Gelman 2000.
12. Nieder 2021.
13. Feigenson, Dehaene, and Spelke 2004.
14. Condry and Spelke 2008.
15. Slusser and Sarnecka 2011.

16. Sarnecka and Carey 2008
17. Condry and Spelke 2008.
18. Wynn 1990.
19. Gelman and Gallistel 1978.
20. Wynn 1990; Sarnecka and Lee 2009.
21. Gordon 2004; Pica et al. 2004.
22. Spaepen et al. 2011; Flaherty and Senghas 2011.
23. Cheung, Rubenson, and Barner 2017.
24. Chu et al. 2020.
25. De Cruz 2006.
26. Tzelgov, Ganor-Stern, and Maymon-Schreiber 2009; Das, LeFevre, and Penner-Wilger 2010.
27. Prather and Alibali 2008.
28. Carey and Barner 2019.
29. Stafylidou and Vosniadou 2004.
30. Moskal and Magone 2000.
31. Braithwaite and Siegler 2018.
32. Vamvakoussi, Van Dooren, and Verschaffel 2012; Varma and Karl 2013.
33. Obersteiner et al. 2013.
34. Alonso-Diaz et al. 2018.
35. Smith, Solomon, and Carey 2005.
36. Smith, Solomon, and Carey 2005.
37. Boyer, Levine, and Huttenlocher 2008.
38. Boyer and Levine 2015; Hurst and Cordes 2018.
39. Moss and Case 1999.
40. Tian, Braithwaite, and Siegler 2021.
41. Siegler et al. 2012.
42. Torbeyns et al. 2015.
43. Strogatz 2013.
44. De Cruz 2009.
45. Goldin et al. 2011
46. Spelke, Lee, and Izard 2010.
47. Dehaene et al. 2006.
48. Dehaene et al. 2006; Izard, Pica, and Spelke 2022.
49. Heimler et al. 2021.
50. Cornell 1985.
51. Logothetis and Pauls 1995.

52. Dehaene et al. 2006; Shusterman, Ah Lee, and Spelke 2008.
53. Dillon, Huang, and Spelke 2013.
54. Dehaene 2010.
55. Barner et al. 2016.

Chapter 7: Ethics

1. Franklin (1791) 2016.
2. Haidt 2003.
3. Bloom 2016.
4. Strohminger and Kumar 2018.
5. Wynn et al. 2018.
6. Bloom 2013.
7. Hofmann et al. 2014.
8. Pennington and Hastie 1986.
9. Piaget (1932) 1965.
10. Cushman et al. 2013; Zelazo, Helwig, and Lau 1996.
11. Cushman et al. 2013.
12. Killen et al. 2011.
13. Moran et al. 2011.
14. Young et al. 2010.
15. McNamara et al. 2019.
16. Martin, Buon, and Cushman 2021.
17. Schmidt, Butler, et al. 2016.
18. Roberts, Gelman, et al. 2017.
19. Roberts, Ho, et al. 2017.
20. Foster-Hanson et al. 2021.
21. Roberts et al. 2018.
22. Tworek and Cimpian 2016.
23. Roberts and Horii 2019.
24. Shtulman and Phillips 2018.
25. Lindstrom et al. 2018.
26. Gachter and Schulz 2016.
27. Bear and Knobe 2017.
28. Bear et al. 2020.
29. Murphy et al. 2009.
30. Kouchaki, Smith, and Savani 2018.
31. Shaw and Olson 2012.

32. Geraci and Surian 2011.
33. Rakoczy, Kaufmann, and Lohse 2016.
34. Baumard, Mascaro, and Chevallier 2012; Rizzo et al. 2016.
35. Paulus, Nöth, and Wörle 2018.
36. Essler and Paulus 2021.
37. Huppert et al. 2019.
38. Schmidt, Svetlova, et al. 2016.
39. Fehr, Bernhard, and Rockenbach 2008.
40. Blake et al. 2015.
41. Starmans, Sheskin, and Bloom 2017.
42. Essler and Paulus 2021.
43. Horne, Powell, and Hummel 2015.
44. Stanley et al. 2018.
45. Lombrozo 2009.
46. Graham et al. 2017.

Chapter 8: Pretense

1. Cray 2020.
2. Singer and Singer 2013.
3. Lillard 2001.
4. Chu and Schulz 2020.
5. Singer et al. 2009.
6. Morelli, Rogoff, and Angelillo 2003.
7. Gaskins 2000.
8. Boyette 2016.
9. Singer et al. 2009.
10. Taggart, Heise, and Lillard 2018.
11. Taggart, Heise, and Lillard 2018; Taggart et al. 2020.
12. Taggart et al. 2020.
13. Gaskins 2000; Morelli, Rogoff, and Angelillo 2003.
14. Lancy 2016.
15. Lew-Levy et al. 2017.
16. Lancy 2016.
17. Nichols and Stitch 2000.
18. Bosco, Friedman, and Leslie 2006.
19. Friedman et al. 2010.
20. Harris, Kavanaugh, and Meredith 1994; Walker-Andrews and Harris 1993.

21. Onishi, Baillargeon, and Leslie 2007.
22. Van de Vondervoort and Friedman 2017.
23. Karmiloff-Smith 1990.
24. Hollis and Low 2005.
25. Ward 1994.
26. Baer and Friedman 2016; Sutherland and Friedman 2013.
27. Hopkins, Dore, and Lillard 2015.
28. Thompson and Goldstein 2020.
29. Sutherland and Friedman 2013.
30. Lillard et al. 2013.
31. Bonawitz et al. 2011; Cook, Goodman, and Schulz 2011.
32. Siegel et al. 2021.
33. Allen and Gutwill 2004.
34. Renken and Nunez 2010.
35. Ganea, Larsen, and Venkadasalam 2021.
36. Miller, Lehman, and Koedinger 1999.
37. Masson, Bub, and Lalonde 2011; Renken and Nunez 2013; Chang, Quintana, and Krajcik 2010; Lewis, Stern, and Linn 1993.
38. Masson, Bub, and Lalonde 2011; Renken and Nunez 2010.
39. Alfieri et al. 2011.
40. Kirschner et al. 2006.
41. Alfieri et al. 2011.
42. Weisberg et al. 2016.
43. Miller et al. 1999.
44. Fisher et al. 2013; Reuter and Leuchter 2020.
45. Rakoczy 2008.
46. Wyman, Rakoczy, and Tomasello 2009.
47. Rakoczy, Warneken, and Tomasello 2008.
48. Riggs and Kalish 2016.
49. Schmidt, Rakoczy, and Tomasello 2012.
50. Schmidt, Rakoczy, et al. 2016.
51. Alex 2020.
52. Wittgenstein 1953.
53. Nguyen 2017.
54. Nguyen 2017; Kapitany, Hampejs, and Goldstein 2022.
55. Dias and Harris 1988.
56. Dias, Roazzi, and Harris 2005.
57. Dias and Harris 1988.

58. Dias and Harris 1990.
59. Leevers and Harris 1999.
60. Richards and Sanderson 1999; Dias, Roazzi, and Harris 2005.
61. Evans, Barston, and Pollard 1983.
62. De Chantal, Gagnon-St-Pierre, and Markovits 2020; De Chantal and Markovits 2022.
63. De Chantal, Gagnon-St-Pierre, and Markovits 2020.
64. Weisberg and Gopnik 2013.
65. Chu and Schulz 2020.

Chapter 9: Fiction

1. Barnes 2012.
2. De Smedt and De Cruz 2015.
3. Nettle 2005.
4. Brockington et al. 2021.
5. Mar and Oatley 2008.
6. Eder 2010.
7. Barnes, Bernstein, and Bloom 2015.
8. Barnes, Bernstein, and Bloom 2015.
9. Weisberg et al. 2013; Sobel and Weisberg 2014.
10. Kibbe, Kreisky, and Weisberg 2018.
11. Thorburn, Bowman-Smith, and Friedman 2020.
12. J.R.R. Tolkien, letter to H. Cotton Minchin, April 16, 1956, quoted in Dubourg and Baumard 2022.
13. Dubourg and Baumard 2022.
14. Weisberg and Goodstein 2009.
15. Gessey-Jones et al. 2020.
16. Stiller, Nettle, and Dunbar 2003.
17. Kelly and Keil 1985.
18. McCoy and Ullman 2019.
19. Lewry et al. 2021.
20. McCoy and Ullman 2019.
21. Shtulman and Morgan 2017; Gong and Shtulman 2021.
22. Shtulman and Morgan 2017.
23. Gong and Shtulman 2021.
24. A. Lane 2006.
25. Norenzayan et al. 2006.

26. Lewis 1978.
27. De Brigard, Henne, and Stanley 2021.
28. Walker, Gopnik, and Ganea 2015.
29. Hopkins and Weisberg 2017.
30. Richert et al. 2009; Richert and Smith 2011.
31. Richert and Smith 2011.
32. Marsh and Fazio 2006; Fazio and March 2008b.
33. Fazio et al. 2012.
34. Fazio and Marsh 2008a.
35. Herbert 1965.
36. Zunshine 2008.
37. Mar et al. 2006.
38. Kynast et al. 2020.
39. Kidd and Costano 2013; Panero et al. 2016; Mar et al. 2006.
40. Dodell-Feder and Tamir 2018.
41. Taylor and Carlson 1997.
42. Roby and Kidd 2008; Trionfi and Reese 2009.
43. Davis, Meins, and Fernyhough 2013.
44. Taylor et al. 2020.
45. Taylor, Cartwright, and Carlson 1993; Davis, Meins, and Fernyhough 2013.
46. Cohen and MacKeith 1991.
47. Del Vecchio et al. 2021.
48. Singh 2021.
49. Appel 2008.
50. Zunshine 2008.
51. Eco 1983.
52. Mares and Acosta 2008.
53. Walker and Lombrozo 2017; Larsen, Lee, and Ganea 2018.
54. Packard 2013.
55. Rottman, Young, and Kelemen 2017; Rottman et al. 2020.
56. Rottman et al. 2020.
57. Gendler 2000.
58. Barnes and Black 2016.
59. Black and Barnes 2020.
60. Shermer 2010.
61. Twain 1897.

Chapter 10: Religion

1. Slone 2004.
2. Johnson et al. 2022.
3. Shtulman and Rattner 2018.
4. Shtulman and Rattner 2018.
5. Shtulman and Rattner 2018.
6. Boyer 2021.
7. Peoples, Duda, and Marlowe 2016.
8. Norenzayan 2013.
9. Guthrie 1993.
10. Lesher 1992.
11. Holtzman 2018.
12. Shtulman and Lindner 2016; Shtulman and Rattner 2018; Richert et al. 2016.
13. Richert et al. 2016; Saide and Richert 2022.
14. Shtulman 2008.
15. Nyhof and Johnson 2017.
16. Shtulman et al. 2019.
17. Barrett and Keil 1996.
18. Barlev, Mermelstein, and German 2017.
19. Barlev, Mermelstein, and German 2018; Barlev et al. 2019.
20. Medina 2022.
21. Heiphetz et al. 2016.
22. Lane, Wellman, and Evans 2010.
23. Lane, Wellman, and Evans 2014.
24. Lane, Evans, et al. 2016.
25. Shtulman and Rattner 2018.
26. Heiphetz et al. 2016.
27. Schjoedt et al. 2009.
28. Purzycki et al. 2012.
29. Epley et al. 2009.
30. Ross, Lelkes, and Russell 2012.
31. Bering 2002.
32. Harris 2011.
33. Bering 2006.
34. Barrett et al. 2021; Astuti and Harris 2008.
35. Lane, Zhu, et al. 2016.
36. Harris 2011.

37. Harris and Gimenez 2005.
38. Harris and Gimenez 2005; Astuti and Harris 2008; Lane, Zhu, et al. 2016.
39. Hodge 2008.
40. White 2016.
41. White 2015.
42. Barlev and Shtulman 2021.
43. Levy 2017.
44. Luhrmann 2020.
45. Van Leeuwen 2014.
46. Shtulman 2013.
47. Clegg et al. 2019; Davoodi et al. 2019; Guerrero, Enesco, and Harris 2010.
48. Cohen, Shariff, and Hill 2008.
49. Liquin, Metz, and Lombrozo 2020; Davoodi and Lombrozo 2022.
50. Heiphetz et al. 2013.
51. Heiphetz, Landers, and Van Leeuwen 2021. The Corpus of Contemporary American English is available at https://www.english-corpora.org/coca/.
52. Heiphetz, Landers, and Van Leeuwen 2021.
53. Van Leeuwen, Weisman, and Luhrmann 2021.
54. Canfield and Ganea 2014.
55. Dore, Jaswal, and Lillard 2015.
56. Oost 1975.
57. Wenger 2001; Watts et al. 2020.
58. Watts et al. 2020.
59. Lupfer, Tolliver, and Jackson 1996; Woolley, Cornelius, and Lacy 2011.
60. Vaden and Woolley 2011; Payir et al. 2021.
61. Payir et al. 2021.
62. Orozco-Giraldo and Harris 2019; Lesage and Richert 2021.
63. Payir et al. 2022.
64. Orozco-Giraldo and Harris 2019.

Chapter 11: Reimagining Imagination

1. Foer 2011.
2. Lucas et al. 2014; Liquin and Gopnik 2022.
3. Liquin and Gopnik 2022.

4. Meder et al. 2021.
5. Phillips and Knobe 2018.
6. Feynman (1964) 2011, pp. 10–11.
7. Mayor 2011.
8. Latson 2017.
9. Thompson, Winn, and Perkins 1979.
10. O'Keane 2021.
11. Dick 1968.
12. Elgammal 2021.
13. L. Lane 2022.
14. GPT-3 2020.
15. Chou and Tversky 2020.
16. Kirk and Lewis 2017.
17. Liberman et al. 2012; Gino and Wiltermuth 2014.
18. Gardner 1993.
19. R. W. Weisberg 1999.
20. Robinson 2007.
21. Said-Metwaly et al. 2021.
22. Lillard 2017.
23. Montessori (1915) 1997.
24. Bilton 2016.

References

Abraham, A., D. Y. Von Cramon, and R. I. Schubotz. 2008. "Meeting George Bush versus Meeting Cinderella: The Neural Response When Telling Apart What Is Real from What Is Fictional in the Context of Our Reality." *Journal of Cognitive Neuroscience* 20: 965–976.

Acierno, J., S. Mischel, and J. Phillips. 2022. "Moral Judgements Reflect Default Representations of Possibility." *Philosophical Transactions of the Royal Society B: Biological Sciences* 377: 20210341.

Adamson, R. E., and D. W. Taylor. 1954. "Functional Fixedness as Related to Elapsed Time and to Set." *Journal of Experimental Psychology* 47: 122–126.

Addis, D. R., Wong, A. T., and Schacter, D. L. 2007. "Remembering the Past and Imagining the Future: Common and Distinct Neural Substrates during Event Construction and Elaboration." *Neuropsychologia* 45: 1363–1377.

Ahl, R. E., D. Amir, and F. C. Keil. 2020. "The World Within: Children Are Sensitive to Internal Complexity Cues." *Journal of Experimental Child Psychology* 200: 104932.

Ahn, W. K., C. W. Kalish, D. L. Medin, and S. A. Gelman. 1995. "The Role of Covariation versus Mechanism Information in Causal Attribution." *Cognition* 54: 299–352.

Alex, B. 2020. "The Ancient History of Board Games." *Discover*, June 16. https://www.discovermagazine.com/planet-earth/the-ancient-history-of-board-games.

Alfieri, L., P. J. Brooks, Aldrich, N. J., and H. R. Tenenbaum. 2011. "Does Discovery-Based Instruction Enhance Learning?" *Journal of Educational Psychology* 103: 1–18.

Allen, S., and J. Gutwill. 2004. "Designing with Multiple Interactives: Five Common Pitfalls." *Curator* 47: 199–212.

Alonso-Díaz, S., Piantadosi, S. T., Hayden, B. Y., and J. F. Cantlon. 2018. "Intrinsic Whole Number Bias in Humans." *Journal of Experimental Psychology: Human Perception and Performance* 44: 1472–1481.

Alvarez, A. L., and A. E. Booth. 2015. "Preschoolers Prefer to Learn Causal Information." *Frontiers in Psychology* 6: 60.

Amsel, E., and S. Brock. 1996. "The Development of Evidence Evaluation Skills." *Cognitive Development* 11: 523–550.

Anderson, C. J., and N. M. Prentice. 1994. "Encounter with Reality: Children's Reactions on Discovering the Santa Claus Myth." *Child Psychiatry and Human Development* 25: 67–84.

Andersson, M. 1982. "Female Choice Selects for Extreme Tail Length in a Widowbird." *Nature* 299: 818–820.

Appel, M. 2008. "Fictional Narratives Cultivate Just-World Beliefs." *Journal of Communication* 58: 62–83.

Astuti, R., and P. L. Harris. 2008. "Understanding Mortality and the Life of the Ancestors in Rural Madagascar." *Cognitive Science* 32: 713–40.

Atance, C. M. 2015. "Young Children's Thinking about the Future." *Child Development Perspectives* 9: 178–182.

Au, T.K.F., C. K. Chan, T. K. Chan, M. W. Cheung, J. Y. Ho, and G. W. Ip. 2008. "Folkbiology Meets Microbiology: A Study of Conceptual and Behavioral Change." *Cognitive Psychology* 57: 1–19.

Baer, C., and O. Friedman. 2016. "Children's Generic Interpretation of Pretense." *Journal of Experimental Child Psychology* 150: 99–111.

Bannister, M. T., A. Bhandare, P. A. Dybczyński, A. Fitzsimmons, A. Guilbert-Lepoutre, R. Jedicke, M. M. Knight, K. J. Meech, A. McNeill, S. Pfalzner, S. N. Raymond, C. Snodgrass, D. E. Trilling, and Q. Ye. 2019. "The Natural History of 'Oumuamua." *Nature Astronomy* 3: 594–602.

Bar, M. 2009. "The Proactive Brain: Memory for Predictions." *Philosophical Transactions of the Royal Society B: Biological Sciences* 364: 1235–1243.

Barlev, M., S. Mermelstein, A. S. Cohen, and T. C. German. 2019. "The Embodied God: Core Intuitions about Person Physicality Coexist and Interfere with Acquired Christian Beliefs about God, the Holy Spirit, and Jesus." *Cognitive Science* 43: e12784.

Barlev, M., S. Mermelstein, and T. C. German. 2017. "Core Intuitions about Persons Coexist and Interfere with Acquired Christian Beliefs about God." *Cognitive Science* 41: 425–454.

———. 2018. "Representational Co-existence in the God Concept: Core Knowledge Intuitions of God as a Person Are Not Revised by Christian Theology despite Lifelong Experience." *Psychonomic Bulletin and Review* 25: 2330–2338.

Barlev, M., and A. Shtulman. 2021. "Minds, Bodies, Spirits, and Gods: Does Widespread Belief in Disembodied Beings Imply That We Are Inherent Dualists?" *Psychological Review* 128: 1007–1021.

Barner, D., G. Alvarez, J. Sullivan, N. Brooks, M. Srinivasan, and M. C. Frank. 2016. "Learning Mathematics in a Visuospatial Format: A Randomized, Controlled Trial of Mental Abacus Instruction." *Child Development* 87: 1146–1158.

Barnes, J. L. 2012. "Fiction, Imagination, and Social Cognition: Insights from Autism." *Poetics* 40: 299–316.

Barnes, J. L., E. Bernstein, and P. Bloom. 2015. "Fact or fiction? Children's Preferences for Real versus Make-Believe Stories." *Imagination, Cognition and Personality* 34: 243–258.

Barnes, J., and J. Black. 2016. "Impossible or Improbable: The Difficulty of Imagining Morally Deviant Worlds." *Imagination, Cognition and Personality* 36: 27–40.

Barrett, H. C., A. Bolyanatz, T. Broesch, E. Cohen, P. Froerer, M. Kanovsky, M. G. Schug, and S. Laurence. 2021. "Intuitive Dualism and Afterlife Beliefs: A Cross-cultural Study." *Cognitive Science* 45: e12992.

Barrett, J. L., and F. C. Keil. 1996. "Conceptualizing a Nonnatural Entity: Anthropomorphism in God Concepts." *Cognitive Psychology* 31: 219–247.

Barrie, J. M. 1911. *Peter and Wendy*. London: Hodder & Stoughton.

Basalla, G. 1988. *The Evolution of Technology*. Cambridge: Cambridge University Press.

Bassok, M., V. M. Chase, and S. A. Martin. 1998. "Adding Apples and Oranges: Alignment of Semantic and Formal Knowledge." *Cognitive Psychology* 35: 99–134.

Bassok, M., and K. J. Holyoak. 1989. "Interdomain Transfer between Isomorphic Topics in Algebra and Physics." *Journal of Experimental Psychology: Learning, Memory, and Cognition* 15: 153–166.

Baumard, N., O. Mascaro, and C. Chevallier. 2012. "Preschoolers Are Able to Take Merit into Account When Distributing Goods." *Developmental Psychology* 48: 492–498.

Bear, A., S. Bensinger, J. Jara-Ettinger, J. Knobe, and F. Cushman. 2020. "What Comes to Mind?" *Cognition* 194: 104057.

Bear, A., and J. Knobe. 2017. "Normality: Part Descriptive, Part Prescriptive." *Cognition* 167: 25–37.

Bearman, C. R., L. J. Ball, and T. C. Ormerod. 2007. "The Structure and Function of Spontaneous Analogising in Domain-based Problem Solving." *Thinking and Reasoning* 13: 273–294.

Beck, S. R., I. A. Apperly, J. Chappell, C. Guthrie, and N. Cutting. 2011. "Making Tools Isn't Child's Play." *Cognition* 119: 301–306.

Beck, S. R., E. J. Robinson, D. J. Carroll, and I. A. Apperly. 2006. "Children's Thinking about Counterfactuals and Future Hypotheticals as Possibilities." *Child Development* 77: 413–426.

Beck, S. R., C. Williams, N. Cutting, I. A. Apperly, and J. Chappell. 2016. "Individual Differences in Children's Innovative Problem-Solving Are Not Predicted by Divergent Thinking or Executive Functions." *Philosophical Transactions of the Royal Society B: Biological Sciences* 371: 20150190.

Bering, J. M. 2002. "Intuitive Conceptions of Dead Agents' Minds: The Natural Foundations of Afterlife Beliefs as Phenomenological Boundary." *Journal of Cognition and Culture* 2: 263–308.

———. 2006. "The Folk Psychology of Souls." *Behavioral and Brain Sciences* 29: 453–498.

Bilalic, M., P. McLeod, and F. Gobet. 2008a. "Inflexibility of Experts—Reality or Myth? Quantifying the Einstellung Effect in Chess Masters." *Cognitive Psychology* 56: 73–102.

———. 2008b. "Why Good Thoughts Block Better Ones: The Mechanism of the Pernicious Einstellung (Set) Effect." *Cognition* 108: 652–661.

Bilton, N. 2016. "How Elizabeth Holmes's House of Cards Came Tumbling Down." *Vanity Fair*, September 6. https://www.vanityfair.com/news/2016/09/elizabeth-holmes-theranos-exclusive.

Black, J. E., and J. L. Barnes. 2019. "Pushing the Boundaries of Reality: Science Fiction, Creativity, and the Moral Imagination." *Psychology of Aesthetics, Creativity, and the Arts* 15: 284–294.

———. 2020. "Morality and the Imagination: Real-World Moral Beliefs Interfere with Imagining Fictional Content." *Philosophical Psychology* 33: 1018–1044.

Black, J. E., S. Capps, and J. Barnes. 2018. "Fiction, Genre Exposure, and Moral Reality." *Psychology of Aesthetics, Creativity, and the Arts* 12: 328–340.

Blair, J. R., J. S. McKee, and L. F. Jernigan. 1980. "Children's Belief in Santa Claus, Easter Bunny and Tooth Fairy." *Psychological Reports* 46: 691–694.

Blake, P. R., K. McAuliffe, J. Corbit, T. C. Callaghan, O. Barry, A. Bowie, L. Kleutsch, K. L. Kramer, E. Ross, H. Vongsachang, R. Wrangham, and F. Warneken. 2015. "The Ontogeny of Fairness in Seven Societies." *Nature* 528: 258–261.

Blancke, S., F. Van Breusegem, G. De Jaeger, J. Braeckman, and M. Van Montagu. 2015. "Fatal Attraction: The Intuitive Appeal of GMO Opposition." *Trends in Plant Science* 20: 414–418.

Blech, C., R. Gaschler, and M. Bilalić. 2020. "Why Do People Fail to See Simple Solutions? Using Think-Aloud Protocols to Uncover the Mechanism behind the Einstellung (Mental Set) Effect." *Thinking and Reasoning* 26: 552–580.

Bloom, P. 2013. *Just Babies: The Origins of Good and Evil*. New York: Crown.

———. 2016. *Against Empathy: The Case for Rational Compassion*. New York: Harper Collins.

Bonawitz, E., P. Shafto, H. Gweon, N. D. Goodman, E. Spelke, and L. E. Schulz. 2011. "The Double-Edged Sword of Pedagogy: Instruction Limits Spontaneous Exploration and Discovery." *Cognition* 120: 322–330.

Bonawitz, E. B., T. J. van Schijndel, D. Friel, and L. E. Schulz. 2012. "Children Balance Theories and Evidence in Exploration, Explanation, and Learning." *Cognitive Psychology* 64: 215–234.

Bonus, J. A., and M. L. Mares. 2019. "Learned and Remembered but Rejected: Preschoolers' Reality Judgments and Transfer from Sesame Street." *Communication Research* 46: 375–400.

Bosco, F. M., O. Friedman, and A. M. Leslie. 2006. "Recognition of Pretend and Real Actions in Play by 1-and 2-Year-Olds: Early Success and Why They Fail." *Cognitive Development* 21: 3–10.

Bowman-Smith, C. K., A. Shtulman, and O. Friedman. 2019. "Distant Lands Make for Distant Possibilities: Children View Improbable Events as More Possible in Far-Away Locations." *Developmental Psychology* 5: 722–728.

Boyer, P. 2021. "Deriving Features of Religions in the Wild." *Human Nature* 32: 557–581.

Boyer, T. W., and S. C. Levine. 2015. "Prompting Children to Reason Proportionally: Processing Discrete Units as Continuous Amounts." *Developmental Psychology* 51: 615–620.

Boyer, T. W., S. C. Levine, and J. Huttenlocher. 2008. "Development of Proportional Reasoning: Where Young Children Go Wrong." *Developmental Psychology* 44: 1478–1490.

Boyette, A. H. 2016. "Children's Play and Culture Learning in an Egalitarian Foraging Society." *Child Development* 87: 759–769.

Braithwaite, D. W., and R. S. Siegler. 2018. "Developmental Changes in the Whole Number Bias." *Developmental Science* 21: e12541.

Brockington, G., A. P. Gomes Moreira, M. S. Buso, S. Gomes da Silva, E. Altszyler, R. Fischer, and J. Moll. 2021. "Storytelling Increases Oxytocin and Positive Emotions and Decreases Cortisol and Pain in Hospitalized Children." *Proceedings of the National Academy of Sciences of the United States of America* 118: e2018409118.

Browne, C. A., and J. D. Woolley. 2004. "Preschoolers' Magical Explanations for Violations of Physical, Social, and Mental Laws." *Journal of Cognition and Development* 5: 239–260.

Buchsbaum, D., S. Bridgers, D. S. Weisberg, and A. Gopnik. 2012. "The Power of Possibility: Causal Learning, Counterfactual Reasoning, and Pretend Play." *Philosophical Transactions of the Royal Society B: Biological Sciences* 367: 2202–2212.

Buchsbaum, D., A. Gopnik, T. L. Griffiths, and P. Shafto. 2011. "Children's Imitation of Causal Action Sequences Is Influenced by Statistical and Pedagogical Evidence." *Cognition* 120: 331–340.

Burdett, E. R., N. McGuigan, R. Harrison, and A. Whiten. 2018. "The Interaction of Social and Perceivable Causal Factors in Shaping Over-imitation." *Cognitive Development* 47: 8–18.

Burnett, F. H. 1911. *The Secret Garden*. 2nd ed. New York: Frederick A. Stokes.

Byrne, R.M.J. 2007. *The Rational Imagination: How People Create Alternatives to Reality*. Cambridge, MA: MIT Press.

Canfield, C. F., and P. A. Ganea. 2014. "'You Could Call It Magic': What Parents and Siblings Tell Preschoolers about Unobservable Entities." *Journal of Cognition and Development* 15: 269–286.

Cantlon, J. F., M. L. Platt, and E. M. Brannon. 2009. "Beyond the Number Domain." *Trends in Cognitive Sciences* 13: 83–91.

Carbonneau, K. J., S. C. Marley, and J. P. Selig. 2013. "A Meta-analysis of the Efficacy of Teaching Mathematics with Concrete Manipulatives." *Journal of Educational Psychology* 105: 380–400.

Carey, S. 2009. *The Origin of Concepts*. Oxford: Oxford University Press.

Carey, S., and D. Barner. 2019. "Ontogenetic Origins of Human Integer Representations." *Trends in Cognitive Sciences* 23: 823–835.

Carr, K., R. L. Kendal, and E. G. Flynn. 2015. "Imitate or Innovate? Children's Innovation Is Influenced by the Efficacy of Observed Behaviour." *Cognition* 142: 322–332.

Casler, K., T. Terziyan, and K. Greene. 2009. "Toddlers View Artifact Function Normatively." *Cognitive Development* 24: 240–247.

Chandler, M. J., and C. E. Lalonde. 1994. "Surprising, Magical and Miraculous Turns of Events: Children's Reactions to Violations of Their Early Theories of Mind and Matter." *British Journal of Developmental Psychology* 12: 83–95.

Chang, H. Y., C. Quintana, and J. S. Krajcik. 2010. "The Impact of Designing and Evaluating Molecular Animations on How Well Middle School Students Understand the Particulate Nature of Matter." *Science Education* 94: 73–94.

Chappell, J., N. Cutting, I. A. Apperly, and S. R. Beck. 2013. The Development of Tool Manufacture in Humans: What Helps Young Children Make Innovative Tools?" *Philosophical Transactions of the Royal Society B: Biological Sciences* 368: 20120409.

Chernyak, N., C. Kang, and T. Kushnir. 2019. "The Cultural Roots of Free Will Beliefs: How Singaporean and US Children Judge and Explain Possibilities for Action in Interpersonal Contexts." *Developmental Psychology* 55: 866–876.

Chernyak, N., T. Kushnir, K. M. Sullivan, and Q. Wang. 2013. "A Comparison of American and Nepalese Children's Concepts of Freedom of Choice and Social Constraint." *Cognitive Science* 37: 1343–1355.

Cheung, P., M. Rubenson, and D. Barner. 2017. "To Infinity and Beyond: Children Generalize the Successor Function to All Possible Numbers Years after Learning to Count." *Cognitive Psychology* 92: 22–36.

Chi, M. T., R. D. Roscoe, J. D. Slotta, M. Roy, and C. C. Chase. 2012. "Misconceived Causal Explanations for Emergent Processes." *Cognitive Science* 36: 1–61.

Chicago Public Media. 2015. "The Incredible Rarity of Changing Your Mind." Episode 555. *This American Life*, April 24. https://www

.thisamericanlife.org/555/the-incredible-rarity-of-changing-your-mind.

Chinn, C. A., and W. F. Brewer. 1998. "An Empirical Test of a Taxonomy of Responses to Anomalous Data in Science. *Journal of Research in Science Teaching* 35: 623–654.

Chou, Y.Y.J., and B. Tversky. 2020. "Changing Perspective: Building Creative Mindsets." *Cognitive Science* 44: e12820.

Chrysikou, E. G., and R. W. Weisberg. 2005. "Following the Wrong Footsteps: Fixation Effects of Pictorial Examples in a Design Problem-Solving Task." *Journal of Experimental Psychology: Learning, Memory, and Cognition* 31: 1134–1148.

Chu, J., P. Cheung, R. M. Schneider, J. Sullivan, and D. Barner. 2020. "Counting to Infinity: Does Learning the Syntax of the Count List Predict Knowledge That Numbers Are Infinite?" *Cognitive Science* 44: e12875.

Chu, J., and L. E. Schulz. 2020. "Play, Curiosity, and Cognition." *Annual Review of Developmental Psychology* 2: 317–343.

Cialone, C., T. Tenbrink, and H. J. Spiers. 2018. "Sculptors, Architects, and Painters Conceive of Depicted Spaces Differently." *Cognitive Science* 42: 524–553.

Clarke, A. C. 1972. *Profiles of the Future: An Inquiry into the Limits of the Possible*. New York: Bantam.

Clay, Z., and C. Tennie. 2018. "Is Overimitation a Uniquely Human Phenomenon? Insights from Human Children as Compared to Bonobos." *Child Development* 89: 1535–1544.

Clayton, N. S., T. J. Bussey, and A. Dickinson. 2003. "Can Animals Recall the Past and Plan for the Future?" *Nature Reviews Neuroscience* 4: 685–691.

Clegg, J. M., Y. K. Cui, P. L. Harris, and K. H. Corriveau. 2019. "God, Germs, and Evolution: Belief in Unobservable Religious and Scientific Entities in the US and China." *Integrative Psychological and Behavioral Science* 53: 93–106.

Clement, J. 1993. "Using Bridging Analogies and Anchoring Intuitions to Deal with Students' Preconceptions in Physics." *Journal of Research in Science Teaching* 30: 1241–1257.

Cohen, A. B., A. F. Shariff, and P. C. Hill. 2008. "The Accessibility of Religious Beliefs." *Journal of Research in Personality* 42: 1408–1417.

Cohen, D., and S. A. MacKeith. 1991. *The Development of Imagination: The Private Worlds of Childhood*. New York: Routledge.

Cook, C., Goodman, N. D., and L. E. Schulz. 2011. "Where Science Starts: Spontaneous Experiments in Preschoolers' Exploratory Play." *Cognition* 120: 341–349.

Cook, C., and D. M. Sobel. 2011. "Children's Beliefs about the Fantasy / Reality Status of Hypothesized Machines." *Developmental Science* 14: 1–8.

Condry, K. F., and E. S. Spelke. 2008. "The Development of Language and Abstract Concepts: The Case of Natural Number." *Journal of Experimental Psychology: General* 137: 22–38.

Cornell, J. M. 1985. "Spontaneous Mirror-Writing in Children." *Canadian Journal of Psychology* 39: 174–179.

Corriveau, K. H., A. L. Kim, C. E. Schwalen, and P. L. Harris. 2009. "Abraham Lincoln and Harry Potter: Children's Differentiation between Historical and Fantasy Characters." *Cognition* 113: 213–225.

Cowen, A. S., and D. Keltner. 2017. "Self-report Captures 27 Distinct Categories of Emotion Bridged by Continuous Gradients." *Proceedings of the National Academy of Sciences of the United States of America* 114: E7900–E7909.

Cray, K. 2020. "How the Coronavirus Is Influencing Children's Play." *The Atlantic*, April 1. https://www.theatlantic.com/family/archive/2020/04/coronavirus-tag-and-other-games-kids-play-during-a-pandemic/609253/.

Cushman, F., R. Sheketoff, S. Wharton, and S. Carey. 2013. "The Development of Intent-based Moral Judgment." *Cognition* 127: 6–21.

Cutting, N., I. A. Apperly, and S. R. Beck. 2011. "Why Do Children Lack the Flexibility to Innovate Tools?" *Journal of Experimental Child Psychology* 109: 497–511.

Cuzzolino, M. P. 2021. "'The Awe Is in the Process': The Nature and Impact of Professional Scientists' Experiences of Awe." *Science Education* 105: 681–706.

Dahl, A., and L. Kim. 2014. "Why Is It Bad to Make a Mess? Preschoolers' Conceptions of Pragmatic Norms." *Cognitive Development* 32: 12–22.

Dahl, R. 1964. *Charlie and the Chocolate Factory*. New York: Alfred A. Knopf.

Danovitch, J. H., and J. D. Lane. 2020. "Children's Belief in Purported Events: When Claims Reference Hearsay, Books, or the Internet." *Journal of Experimental Child Psychology* 193: 104808.

Darwin, C. 1871. *The Descent of Man, and Selection in Relation to Sex.* 2 vols. London: John Murray.

Das, R., J. A. LeFevre, and M. Penner-Wilger. 2010. "Negative Numbers in Simple Arithmetic." *Quarterly Journal of Experimental Psychology* 63: 1943–1952.

Davis, P. E., E. Meins, and C. Fernyhough. 2013. "Individual Differences in Children's Private Speech: The Role of Imaginary Companions." *Journal of Experimental Child Psychology* 116: 561–571.

Davoodi, T., M. Jamshidi-Sianaki, F. Abedi, A. Payir, Y. K. Cui, Harris, P. L., and K. H. Corriveau. 2019. "Beliefs about Religious and Scientific Entities among Parents and Children in Iran." *Social Psychological and Personality Science* 10: 847–855.

Davoodi, T., and T. Lombrozo. 2022. "Varieties of Ignorance: Mystery and the Unknown in Science and Religion." *Cognitive Science* 46: e13129.

Dawkins, R. 1993. "Viruses of the Mind." In *Dennett and His Critics: Demystifying Mind*, edited by B. Dahlbom, 13–27. Cambridge, MA: Blackwell.

De Brigard, F., P. Henne, and M. L. Stanley. 2021. "Perceived Similarity of Imagined Possible Worlds Affects Judgments of Counterfactual Plausibility." *Cognition* 209: 104574.

De Brigard, F., and N. Parikh. 2019. "Episodic Counterfactual Thinking." *Current Directions in Psychological Science* 28: 59–66.

De Chantal, P. L., É. Gagnon-St-Pierre, and H. Markovits. 2020. "Divergent Thinking Promotes Deductive Reasoning in Preschoolers." *Child Development* 91: 1081–1097.

De Chantal, P. L., and H. Markovits. 2022. "Reasoning outside the Box: Divergent Thinking Is Related to Logical Reasoning." *Cognition* 224: 105064.

De Cruz, H. 2006. "Why Are Some Numerical Concepts More Successful Than Others? An Evolutionary Perspective on the History of Number Concepts." *Evolution and Human Behavior* 27: 306–323.

———. 2009. "An Enhanced Argument for Innate Elementary Geometric Knowledge and Its Philosophical Implications." In *New Perspectives on Mathematical Practices: Essays in Philosophy and History of Mathematics*, edited by B. Van Kerkhove, 185–206. Hackensack, NJ: World Scientific.

De Cruz, H., and J. De Smedt. 2010. "Science as Structured Imagination." *Journal of Creative Behavior* 44: 37–52.

Defeyter, M. A., and T. P. German. 2003. "Acquiring an Understanding of Design: Evidence from Children's Insight Problem Solving." *Cognition* 89: 133–155.

Dehaene, S. 2010. *Reading in the Brain: The New Science of How We Read*. New York: Penguin.

Dehaene, S., V. Izard, P. Pica, and E. Spelke. 2006. "Core Knowledge of Geometry in an Amazonian Indigene Group." *Science* 311: 381–384.

Del Vecchio, M., A. Kharlamov, G. Parry, and G. Pogrebna. 2021. "Improving Productivity in Hollywood with Data Science: Using Emotional Arcs of Movies to Drive Product and Service Innovation in Entertainment Industries." *Journal of the Operational Research Society* 72: 1110–1137.

Dennett, D. C. 1993. *Consciousness Explained*. New York: Penguin.

De Saint-Exupery, A. 1970. *The Little Prince*. Translated by Katherine Woods. New York: Harcourt Brace & World. Original work published in 1943.

De Smedt, J., and H. De Cruz. 2015. "The Epistemic Value of Speculative Fiction." *Midwest Studies in Philosophy* 39: 58–77.

Dias, M. G., and P. L. Harris. 1988. "The Effect of Make-Believe Play on Deductive Reasoning." *British Journal of Developmental Psychology* 6: 207–221.

———. 1990. "The Influence of the Imagination on Reasoning by Young Children." *British Journal of Developmental Psychology* 8: 305–318.

Dias, M. G., A. Roazzi, and P. L. Harris. 2005. "Reasoning from Unfamiliar Premises: A Study with Unschooled Adults." *Psychological Science* 16: 550–554.

Dick, P. K. 1968. *Do Androids Dream of Electric Sheep?* New York: Ballantine.

Dillon, M. R., Y. Huang, and E. S. Spelke. 2013. "Core Foundations of Abstract Geometry." *Proceedings of the National Academy of Sciences of the United States of America* 110: 14191–14195.

DiYanni, C., and D. Kelemen. 2008. "Using a Bad Tool with Good Intention: Young Children's Imitation of Adults' Questionable Choices." *Journal of Experimental Child Psychology* 101: 241–261.

Dodell-Feder, D., and D. I. Tamir. 2018. "Fiction Reading Has a Small Positive Impact on Social Cognition: A Meta-analysis." *Journal of Experimental Psychology: General* 147: 1713–1727.

Dore, R. A., V. K. Jaswal, and A. S. Lillard. 2015. "Real or Not? Informativeness Influences Children's Reality Status Judgments." *Cognitive Development* 33: 28–39.

Dubourg, E., and N. Baumard. 2022. "Why Imaginary Worlds?: The Psychological Foundations and Cultural Evolution of Fictions with Imaginary Worlds." *Behavioral and Brain Sciences* 45: e276.

Dumas, D., P. A. Alexander, L. M. Baker, S. Jablansky, and K. N. Dunbar. 2014. "Relational Reasoning in Medical Education: Patterns in Discourse and Diagnosis." *Journal of Educational Psychology* 106: 1021–1035.

Dunbar, K., and I. Blanchette. 2001. "The in Vivo / in Vitro Approach to Cognition: The Case of Analogy." *Trends in Cognitive Sciences* 5: 334–339.

Duncker, K. 1945. "On Problem-Solving." *Psychological Monographs* 58: i–113.

Eco, U. 1983. *The Name of the Rose*. New York: Harcourt.

Eder, J. 2010. "Understanding Characters." *Projections* 4: 16–40.

Edwards, B. J., J. J. Williams, D. Gentner, and T. Lombrozo. 2019. "Explanation Recruits Comparison in a Category-Learning Task." *Cognition* 185: 21–38.

Elgammal, A. 2021. "How Artificial Intelligence Completed Beethoven's Unfinished Tenth Symphony." *Smithsonian Magazine*, September 24. https://www.smithsonianmag.com/innovation/how-artificial-intelligence-completed-beethovens-unfinished-10th-symphony-180978753/.

Engle, J., and C. M. Walker. 2021. "Thinking Counterfactually Supports Children's Evidence Evaluation in Causal Learning." *Child Development* 92: 1636–1651.

Epley, N., B. A. Converse, A. Delbosc, G. A. Monteleone, and J. T. Cacioppo. 2009. "Believers' Estimates of God's Beliefs Are More Egocentric Than Estimates of Other People's Beliefs." *Proceedings of the National Academy of Sciences of the United States of America* 106: 21533–21538.

Essler, S., and M. Paulus. 2021. "Robin Hood or Matthew? Children's Reasoning about Redistributive Justice in the Context of Economic Inequalities." *Child Development* 92: 1254–1273.

Evans, J. St. B. T., J. L. Barston, and P. Pollard. 1983. "On the Conflict between Logic and Belief in Syllogistic Reasoning." *Memory and Cognition* 11: 295–306.

Fazio, L. K., S. J. Barber, Rajaram, S., P. A. Ornstein, and E. J. Marsh. 2012. "Creating Illusions of Knowledge: Learning Errors That Contradict Prior Knowledge." *Journal of Experimental Psychology: General* 142: 1–5.

Fazio, L. K., and E. J. Marsh. 2008a. "Older, Not Younger, Children Learn More False Facts from Stories." *Cognition* 106: 1081–1089.

———. 2008b. "Slowing Presentation Speed Increases Illusions of Knowledge." *Psychonomic Bulletin and Review* 15: 180–185.

Fehr, E., H. Bernhard, and B. Rockenbach. 2008. "Egalitarianism in Young Children." *Nature* 454: 1079–1083.

Feigenson, L., S. Dehaene, and E. Spelke. 2004. "Core Systems of Number." *Trends in Cognitive Sciences* 8: 307–314.

Feynman, R. P. (1964) 2011. *The Feynman Lectures on Physics*. Vol. 2. New York: Basic Books.

Fisher, K. R., K. Hirsh-Pasek, N. Newcombe, and R. M. Golinkoff. 2013. "Taking Shape: Supporting Preschoolers' Acquisition of Geometric Knowledge through Guided Play." *Child Development* 84: 1872–1878.

Fisher, M., and F. C. Keil. 2016. "The Curse of Expertise: When More Knowledge Leads to Miscalibrated Explanatory Insight." *Cognitive Science* 40: 1251–1269.

Fitch, W. T., and J. Giedd. 1999. "Morphology and Development of the Human Vocal Tract: A Study Using Magnetic Resonance Imaging." *Journal of the Acoustical Society of America* 106: 1511–1522.

FitzGibbon, L., A. Komiya, and K. Murayama. 2021. "The Lure of Counterfactual Curiosity: People Incur a Cost to Experience Regret." *Psychological Science* 32: 241–255.

Flaherty, M., and A. Senghas. 2011. "Numerosity and Number Signs in Deaf Nicaraguan Adults." *Cognition* 121: 427–436.

Flynn, E. 2008. "Investigating Children as Cultural Magnets: Do Young Children Transmit Redundant Information along Diffusion Chains?" *Philosophical Transactions of the Royal Society B: Biological Sciences* 363: 3541–3551.

Foer, J. 2011. *Moonwalking with Einstein: The Art and Science of Remembering Everything*. New York: Penguin.

Foster, M. I., and M. T. Keane. 2015. "Why Some Surprises Are More Surprising Than Others: Surprise as a Metacognitive Sense of Explanatory Difficulty." *Cognitive Psychology* 81: 74–116.

Foster-Hanson, E., S. O. Roberts, S. A. Gelman, and M. Rhodes. 2021. "Categories Convey Prescriptive Information across Domains and Development." *Journal of Experimental Child Psychology* 212: 105231.

Franklin, B. (1791) 2016. *Book of Virtues*. Carlisle, MA: Applewood Books.

Frazier, B. N., S. A. Gelman, and H. M. Wellman. 2016. "Young Children Prefer and Remember Satisfying Explanations." *Journal of Cognition and Development* 17: 718–736.

Friedman, O., K. R. Neary, C. L. Burnstein, and A. M. Leslie. 2010. "Is Young Children's Recognition of Pretense Metarepresentational or Merely Behavioral? Evidence from 2-and 3-Year-Olds' Understanding of Pretend Sounds and Speech." *Cognition* 115: 314–319.

Fugelsang, J. A., and K. N. Dunbar. 2005. "Brain-based Mechanisms Underlying Complex Causal Thinking." *Neuropsychologia* 43: 1204–1213.

Fyfe, E. R., N. M. McNeil, J. Y. Son, and R. L. Goldstone. 2014. "Concreteness Fading in Mathematics and Science Instruction: A Systematic Review." *Educational Psychology Review* 26: 9–25.

Gachter, S., and J. F. Schulz. 2016. "Intrinsic Honesty and the Prevalence of Rule Violations across Societies." *Nature* 531: 496–499.

Gallistel, C. R., and R. Gelman. 2000. "Non-verbal Numerical Cognition: From Reals to Integers." *Trends in Cognitive Sciences* 4: 59–65.

Ganea, P. A., N. E. Larsen, and V. P. Venkadasalam. 2021. "The Role of Alternative Theories and Anomalous Evidence in Children's Scientific Belief Revision." *Child Development* 92: 1137–1153.

Gamarra, N. T. 1969. *Erroneous Predictions and Negative Comments Concerning Scientific and Technological Developments*. Congressional Research Report CB-150, F-381. Washington, DC: Congressional Research Service. https://digital.library.unt.edu/ark:/67531/metadc1038944/.

Gardiner, A. K., D. F. Bjorklund, M. L. Greif, and S. K. Gray. 2012. "Choosing and Using Tools: Prior Experience and Task Difficulty Influence Preschoolers' Tool-use Strategies." *Cognitive Development* 27: 240–254.

Gardner, H. 1993. *Creating Minds: An Anatomy of Creativity Seen through the Lives of Freud, Einstein, Picasso, Stravinsky, Eliot, Graham, and Gandhi*. New York: Basic Books.

Gaskins, S. 2000. "Children's Daily Activities in a Mayan Village: A Culturally Grounded Description." *Cross-Cultural Research* 34: 375–389.

Gelman, R., and C. R. Gallistel. 1978. *The Child's Understanding of Number*. Cambridge, MA: Harvard University Press.

Gendler, T. S. 2000. "The Puzzle of Imaginative Resistance." *Journal of Philosophy* 97: 55–81.

Gentner, D. 1983. "Structure-Mapping: A Theoretical Framework for Analogy." *Cognitive Science* 7: 155–170.

Gentner, D., S. Brem, R. W. Ferguson, A. B. Markman, B. B. Levidow, P. Wolff, and K. D. Forbus. 1997. "Analogical Reasoning and Conceptual Change: A Case Study of Johannes Kepler." *Journal of the Learning Sciences* 6: 3–40.

Gentner, D., J. Loewenstein, and L. Thompson. 2003. "Learning and Transfer: A General Role for Analogical Encoding." *Journal of Educational Psychology* 95: 393–408.

Gentner, D., and C. Toupin. 1986. "Systematicity and Surface Similarity in the Development of Analogy." *Cognitive Science* 10: 277–300.

Geraci, A., and L. Surian. 2011. "The Developmental Roots of Fairness: Infants' Reactions to Equal and Unequal Distributions of Resources." *Developmental Science* 14: 1012–1020.

Gergely, G., and G. Csibra. 2006. "Sylvia's Recipe: The Role of Imitation and Pedagogy in the Transmission of Cultural Knowledge." In *Roots of Human Sociality: Culture, Cognition and Interaction*, edited by N. J. Enfield and S. C. Levinson, 229–255. New York: Routledge.

German, T. P. 1999. "Children's Causal Reasoning: Counterfactual Thinking Occurs for Negative Outcomes Only." *Developmental Science* 2: 442–457.

German, T. P., and H. C. Barrett. 2005. "Functional Fixedness in a Technologically Sparse Culture." *Psychological Science* 16: 1–5.

German, T. P., and M. A. Defeyter. 2000. "Immunity to Functional Fixedness in Young Children." *Psychonomic Bulletin and Review* 7: 707–712.

Gessey-Jones, T., C. Connaughton, R. Dunbar, R. Kenna, P. MacCarron, C. O'Conchobhair, and J. Yose. 2020. "Narrative Structure of *A Song of Ice and Fire* Creates a Fictional World with Realistic Measures of Social Complexity." *Proceedings of the National Academy of Sciences of the United States of America* 117: 28582–28588.

Gino, F., and S. S. Wiltermuth. 2014. "Evil Genius? How Dishonesty Can Lead to Greater Creativity." *Psychological Science* 25: 973–981.

Glucksberg, S.A.M., and J. H. Danks. 1968. "Effects of Discriminative Labels and of Nonsense Labels upon Availability of Novel Function." *Journal of Memory and Language* 7: 72–76.

Goldin, A. P., L. Pezzatti, A. M. Battro, and M. Sigman. 2011. "From Ancient Greece to Modern Education: Universality and Lack of Generalization of the Socratic Dialogue." *Mind, Brain, and Education* 5: 180–185.

Goldstein, T. R., and J. Woolley. 2016. "Ho! Ho! Who? Parent Promotion of Belief in and Live Encounters with Santa Claus." *Cognitive Development* 39: 113–127.

Goldwater, M. B., H. J. Don, M. J. Krusche, and E. J. Livesey. 2018. "Relational Discovery in Category Learning." *Journal of Experimental Psychology: General* 147: 1–35.

Goldwater, M. B., and D. Gentner. 2015. "On the Acquisition of Abstract Knowledge: Structural Alignment and Explication in Learning Causal System Categories." *Cognition* 137: 137–153.

Goldwater, M. B., D. Gentner, N. D. LaDue, and J. C. Libarkin. 2021. "Analogy Generation in Science Experts and Novices." *Cognitive Science* 45: e13036.

Gong, T., and A. Shtulman. 2021. "The Plausible Impossible: Chinese Adults Hold Graded Notions of Impossibility." *Journal of Cognition and Culture* 21: 76–93.

Gonul, G., A. Hohenberger, M. Corballis, and A. M. Henderson. 2019. "Joint and Individual Tool Making in Preschoolers: From Social to Cognitive Processes." *Social Development* 28: 1037–1053.

Goodwin, C. 1994. "Professional Vision." *American Anthropologist* 96: 606–633.

Gopnik, A. 2009. *The Philosophical Baby: What Children's Minds Tell Us about Truth, Love, and the Meaning of Life*. New York: Farrar, Straus and Giroux.

———. 2016. *The Gardener and the Carpenter: What the New Science of Child Development Tells Us about the Relationship between Parents and Children*. New York: Farrar, Straus and Giroux.

Gordon, P. 2004. "Numerical Cognition without Words: Evidence from Amazonia." *Science* 306: 496–499.

Gottlieb, S., D. Keltner, and T. Lombrozo. 2018. "Awe as a Scientific Emotion." *Cognitive Science* 42: 2081–2094.

Gould, S. J. 1992. *Ever since Darwin: Reflections in Natural History*. New York: W.W. Norton.

Goulding, B. W., and O. Friedman. 2020. "Children's Beliefs about Possibility Differ across Dreams, Stories, and Reality." *Child Development* 91: 1843–1853.

Goulding, B. W., E. E. Stonehouse, and O. Friedman. 2022. "Causal Knowledge and Children's Possibility Judgments." *Child Development* 93: 794–803.

GPT-3. 2020. "A Robot Wrote This Entire Article. Are You Scared Yet, Human?" *The Guardian*, September 8. https://www.theguardian.com/commentisfree/2020/sep/08/robot-wrote-this-article-gpt-3.

Graham, J., A. Waytz, P. Meindl, R. Iyer, and L. Young. 2017. "Centripetal and Centrifugal Forces in the Moral Circle: Competing Constraints on Moral Learning." *Cognition* 167: 58–65.

Gregg, V. R., G. A. Winer, J. E. Cottrell, K. E. Hedman, and J. S. Fournier. 2001. "The Persistence of a Misconception about Vision after Educational Interventions." *Psychonomic Bulletin and Review* 8: 622–626.

Gregory, T. R. 2009. "Understanding Natural Selection: Essential Concepts and Common Misconceptions." *Evolution: Education and Outreach* 2: 156–175.

Gropen, J., N. Clark-Chiarelli, C. Hoisington, and S. B. Ehrlich. 2011. "The Importance of Executive Function in Early Science Education." *Child Development Perspectives* 5: 298–304.

Gros, H., E. Sander, and J. P. Thibaut. 2019. "When Masters of Abstraction Run into a Concrete Wall: Experts Failing Arithmetic Word Problems." *Psychonomic Bulletin and Review* 26: 1738–1746.

Grotzer, T. A., S. L. Solis, M. S. Tutwiler, and M. P. Cuzzolino. 2017. "A Study of Students' Reasoning about Probabilistic Causality: Implications for Understanding Complex Systems and for Instructional Design." *Instructional Science* 45: 25–52.

Guerrero, S., I. Enesco, and P. L. Harris. 2010. "Oxygen and the Soul: Children's Conception of Invisible Entities." *Journal of Cognition and Culture* 10: 123–151.

Guthormsen, A. M., K. J. Fisher, M. Bassok, L. Osterhout, M. DeWolf, and K. J. Holyoak. 2016. "Conceptual Integration of Arithmetic Operations with Real-World Knowledge: Evidence from Event-related Potentials." *Cognitive Science* 40: 723–757.

Guthrie, S. 1993. *Faces in the Clouds: A New Theory of Religion*. Oxford: Oxford University Press.

Haidt, J. 2003. "The Moral Emotions." In *Handbook of Affective Sciences*, edited by R. J. Davidson, K. R. Scherer, and H. H. Goldsmith, 852–870. New York: Oxford University Press.

Harris, P. L. 2011. "Conflicting Thoughts about Death." *Human Development* 54: 160–168.

Harris, P. L., T. German, and P. Mills. 1996. "Children's Use of Counterfactual Thinking in Causal Reasoning." *Cognition* 61: 233–259.

Harris, P. L. and M. Gimenez. 2005. "Children's Acceptance of Conflicting Testimony: The Case of Death." *Journal of Cognition and Culture* 5: 143–164.

Harris, P. L., R. D. Kavanaugh, and M. C. Meredith. 1994. "Young Children's Comprehension of Pretend Episodes: The Integration of Successive Actions." *Child Development* 65: 16–30.

Hassabis, D., D. Kumaran, S. D. Vann, and E. A. Maguire. 2007. "Patients with Hippocampal Amnesia Cannot Imagine New Experiences." *Proceedings of the National Academy of Sciences of the United States of America* 104: 1726–1731.

Hauser, M., F. Cushman, L. Young, R. K. Jin, and J. Mikhail. 2007. "A Dissociation between Moral Judgments and Justifications." *Mind and Language* 22: 1–21.

Heimler, B., T. Behor, S. Dehaene, V. Izard, and A. Amedi. 2021. "Core Knowledge of Geometry Can Develop Independently of Visual Experience." *Cognition* 212: 104716.

Heiphetz, L., C. L. Landers, and N. Van Leeuwen. 2021. "Does Think Mean the Same Thing as Believe? Linguistic Insights into Religious Cognition." *Psychology of Religion and Spirituality* 13: 287–297.

Heiphetz, L., J. D. Lane, A. Waytz, and L. L. Young. 2016. "How Children and Adults Represent God's Mind." *Cognitive Science* 40: 121–144.

Heiphetz, L., E. S. Spelke, P. L. Harris, and M. R. Banaji. 2013. "The Development of Reasoning about Beliefs: Fact, Preference, and Ideology." *Journal of Experimental Social Psychology* 49: 559–565.

Hemmerich, J. A., K. Van Voorhis, and J. Wiley. 2016. "Anomalous Evidence, Confidence Change, and Theory Change." *Cognitive Science* 40: 1534–1560.

Henderson, M. D., and C. J. Wakslak. 2010. "Over the Hills and Far Away: The Link between Physical Distance and Abstraction." *Current Directions in Psychological Science* 19: 390–394.

Henrich, J. 2016. *The Secret of Our Success: How Culture Is Driving Human Evolution, Domesticating Our Species, and Making Us Smarter.* Princeton, NJ: Princeton University Press.

Herbert, F. 1965. *Dune*. Sudbury, United Kingdom: Chilton Books.

Hickling, A. K., and H. M. Wellman. 2001. "The Emergence of Children's Causal Explanations and Theories: Evidence from Everyday Conversation." *Developmental Psychology* 37: 668–683.

Hilbrink, E. E., E. Sakkalou, K. Ellis-Davies, N. C. Fowler, and M. Gattis. 2013. "Selective and Faithful Imitation at 12 and 15 Months." *Developmental Science* 16: 828–840.

Hobsbawm, E. J. 1952. "The Machine Breakers." *Past and Present* 1: 57–70.

Hodge, K. M. 2008. "Descartes' Mistake: How Afterlife Beliefs Challenge the Assumption That Humans Are Intuitive Cartesian Substance Dualists." *Journal of Cognition and Culture* 8: 387–415.

Hofmann, W., D. C. Wisneski, M. J. Brandt, and L. J. Skitka. 2014. "Morality in Everyday Life." *Science* 345: 1340–1343.

Hofstadter, D. R., and E. Sander. 2013. *Surfaces and Essences: Analogy as the Fuel and Fire of Thinking.* New York: Basic Books.

Holden, M. P., N. S. Newcombe, I. Resnick, and T. F. Shipley. 2016. "Seeing Like a Geologist: Bayesian Use of Expert Categories in Location Memory." *Cognitive Science* 40: 440–454.

Hollis, S., and J. Low. 2005. "Karmiloff-Smith's RRM Distinction between Adjunctions and Redescriptions: It's about Time (and Children's Drawings)." *British Journal of Developmental Psychology* 23: 623–644.

Holmes, R. 2010. *The Age of Wonder: How the Romantic Generation Discovered the Beauty and Terror of Science.* New York: Vintage.

Holtzman, L. 2018. *Anthropomorphism in Islam.* Edinburgh: Edinburgh University Press.

Hoover, J. D., and A. F. Healy. 2017. "Algebraic Reasoning and Bat-and-Ball Problem Variants: Solving Isomorphic Algebra First Facilitates Problem Solving Later." *Psychonomic Bulletin and Review* 24: 1922–1928.

Horner, V., and A. Whiten. 2005. "Causal Knowledge and Imitation / Emulation Switching in Chimpanzees (*Pan troglodytes*) and Children (*Homo sapiens*)." *Animal Cognition* 8: 164–181.

Hopkins, E. J., R. A. Dore, and A. S. Lillard. 2015. "Do Children Learn from Pretense?" *Journal of Experimental Child Psychology* 130: 1–18.

Hopkins, E. J., and D. S. Weisberg. 2017. "The Youngest Readers' Dilemma: A Review of Children's Learning from Fictional Sources." *Developmental Review* 43: 48–70.

Horne, Z., D. Powell, and J. Hummel. 2015. "A Single Counterexample Leads to Moral Belief Revision." *Cognitive Science* 39: 1950–1964.

Hsee, C. K., and B. Ruan. 2016. "The Pandora Effect: The Power and Peril of Curiosity." *Psychological Science* 27: 659–666.

Hughes, H. C. 1999. *Sensory Exotica: A World beyond Human Experience*. Cambridge, MA: MIT Press.

Huppert, E., J. M. Cowell, Y. Cheng, C. Contreras-Ibáñez, N. Gomez-Sicard, M. L. Gonzalez-Gadea, D. Huepe, A. Ibanez, K. Lee, R. Mahasneh, S. Malcolm-Smith, N. Salas, B. Selcuk, B. Tungodden, A. Wong, X. Zhou, and J. Decety. 2019. "The Development of Children's Preferences for Equality and Equity across 13 Individualistic and Collectivist Cultures." *Developmental Science* 22: e12729.

Hurst, M. A., and S. Cordes. 2018. "Attending to Relations: Proportional Reasoning in 3-to 6-Year-Old Children." *Developmental Psychology* 54: 428–439.

Hutson, M. 2020. "Eye-catching Advances in Some AI Fields Are Not Real." *Science*, May 27. https://doi.org/10.1126/science.abd0313.

Irish, M., and O. Piguet. 2013. "The Pivotal Role of Semantic Memory in Remembering the Past and Imagining the Future." *Frontiers in Behavioral Neuroscience* 7: 27.

Islam, M. S., A.H.M. Kamal, A. Kabir, D. L. Southern, S. H. Khan, S. M. Hasan, T. Sarkar, S. Sharmin, S. Das, T. Roy, M.G.D. Harun, A. A. Chughtai, N. Homaira, and H. Seale. 2021. "COVID-19 Vaccine Rumors and Conspiracy Theories: The Need for Cognitive Inoculation against Misinformation to Improve Vaccine Adherence." *PLoS One* 16: e0251605.

Izard, V., P. Pica, and E. S. Spelke. 2022. "Visual Foundations of Euclidean Geometry." *Cognitive Psychology* 136: 101494.

Jamrozik, A., and D. Gentner. 2020. "Relational Labeling Unlocks Inert Knowledge." *Cognition* 196: 104146.

Jansson, D. G., and S. M. Smith. 1991. "Design Fixation." *Design Studies* 12: 3–11.

Jee, B. D., and F. K. Anggoro. 2021. "Designing Exhibits to Support Relational Learning in a Science Museum." *Frontiers in Psychology* 12: 636030.

Jee, B. D., D. H. Uttal, D. Gentner, C. Manduca, T. F. Shipley, and B. Sageman. 2013. "Finding Faults: Analogical Comparison Supports Spatial Concept Learning in Geoscience." *Cognitive Processing* 14: 175–187.

Johnson, C. N., and P. L. Harris. 1994. "Magic: Special but Not Excluded." *British Journal of Developmental Psychology* 12: 35–51.

Johnson, K. A., A. B. Weinberger, E. Dyke, G. F. Porter, D. J. Kraemer, J. Grafman, A. B. Cohen and A. E. Green. 2022. "Differentiating Personified, Supernatural, and Abstract Views of God across Three Cognitive Domains." *Psychology of Religion and Spirituality*, May 2022. https://psycnet.apa.org/doi/10.1037/rel0000460.

Johnson, S. 2002. *Emergence: The Connected Lives of Ants, Brains, Cities, and Software*. New York: Simon & Schuster.

———. 2014. *How We Got to Know: Six Innovations That Made the Modern World*. New York: Riverhead.

———. 2021. *Extra Life: A Short History of Living Longer*. New York: Riverhead.

Johnson, S. G., and S. Steinerberger. 2019. "Intuitions about Mathematical Beauty: A Case Study in the Aesthetic Experience of Ideas." *Cognition* 189: 242–259.

Johnson-Laird, P. N. 2005. "Flying Bicycles: How the Wright Brothers Invented the Airplane." *Mind and Society* 4: 27–48.

Johnston, A. M., P. C. Holden, and L. R. Santos. 2017. "Exploring the Evolutionary Origins of Overimitation: A Comparison across Domesticated and Non-domesticated Canids." *Developmental Science* 20: e12460.

Joslyn, S., and R. Demnitz. 2021. "Explaining How Long CO_2 Stays in the Atmosphere: Does It Change Attitudes toward Climate Change?" *Journal of Experimental Psychology: Applied* 27: 473–484.

Juma, C. 2016. *Innovation and Its Enemies: Why People Resist New Technologies*. New York: Oxford University Press.

Kalish, C. 1998. "Reasons and Causes: Children's Understanding of Conformity to Social Rules and Physical Laws." *Child Development* 69: 706–720.

Kapitany, R., T. Hampejs, and T. R. Goldstein. 2022. "Pretensive Shared Reality: From Childhood Pretense to Adult Imaginative Play." *Frontiers in Psychology* 13: 774085.

Karmiloff-Smith, A. 1990. "Constraints on Representational Change: Evidence from Children's Drawing." *Cognition* 34: 57–83.

Keil, F. C. 2022. *Wonder: Childhood and the Lifelong Love of Science*. Cambridge, MA: MIT Press.

Keil, F. C., and K. L. Lockhart. 2021. "Beyond Cause: The Development of Clockwork Cognition." *Current Directions in Psychological Science* 30: 167–173.

Kelemen, D. 2019. "The Magic of Mechanism: Explanation-based Instruction on Counterintuitive Concepts in Early Childhood." *Perspectives on Psychological Science* 14: 510–522.

Kelemen, D., N. A. Emmons, R. Seston Schillaci, and P. A. Ganea. 2014. "Young Children Can Be Taught Basic Natural Selection Using a Picture-Storybook Intervention." *Psychological Science* 25: 893–902.

Kelly, M. H., and F. C. Keil. 1985. "The More Things Change . . . : Metamorphoses and Conceptual Structure." *Cognitive Science* 9: 403–416.

Keltner, D., and J. Haidt. 2003. "Approaching Awe, a Moral, Spiritual, and Aesthetic Emotion." *Cognition and Emotion* 17: 297–314.

Kenward, B. 2012. "Over-imitating Preschoolers Believe Unnecessary Actions Are Normative and Enforce Their Performance by a Third Party." *Journal of Experimental Child Psychology* 112: 195–207.

Kenward, B., M. Karlsson, and J. Persson. 2011. "Over-imitation Is Better Explained by Norm Learning Than by Distorted Causal Learning." *Proceedings of the Royal Society B: Biological Sciences* 278: 1239–1246.

Keupp, S., T. Behne, and H. Rakoczy. 2013. "Why Do Children Overimitate? Normativity Is Crucial." *Journal of Experimental Child Psychology* 116: 392–406.

Kibbe, M. M., M. Kreisky, and D. S. Weisberg. 2018. "Young Children Distinguish between Different Unrealistic Fictional Genres." *Psychology of Aesthetics, Creativity, and the Arts* 12: 228–235.

Kidd, C., and B. Y. Hayden. 2015. "The Psychology and Neuroscience of Curiosity." *Neuron* 88: 449–460.

Kidd, D. C., and E. Castano. 2013. "Reading Literary Fiction Improves Theory of Mind." *Science* 342: 377–380.

Kiley, K., and S. Vaisey. 2020. "Measuring Stability and Change in Personal Culture Using Panel Data." *American Sociological Review* 85: 477–506.

Killen, M., K. L. Mulvey, C. Richardson, N. Jampol, and A. Woodward. 2011. "The Accidental Transgressor: Morally-Relevant Theory of Mind." *Cognition* 119: 197–215.

Kirk, E., and C. Lewis. 2017. "Gesture Facilitates Children's Creative Thinking." *Psychological Science* 28: 225–232.

Kirschner, P. A., J. Sweller, and R. E. Clark. 2006. "Why Minimal Guidance during Instruction Does Not Work: An Analysis of the Failure of

Constructivist, Discovery, Problem-based, Experiential, and Inquiry-based Teaching." *Educational Psychologist* 41: 75–86.

Klahr, D., and Z. Chen. 2003. "Overcoming the Positive-Capture Strategy in Young Children: Learning about Indeterminacy." *Child Development* 74: 1275–1296.

———. 2011. "Finding One's Place in Transfer Space." *Child Development Perspectives* 5: 196–204.

Kline, M. A., M. M. Gervais, C. Moya, and R. T. Boyd. 2020. "Irrelevant-Action Imitation Is Short-term and Contextual: Evidence from Two Under-studied Populations." *Developmental Science* 23: e12903.

Klotz, I. M. 1980. "The N-Ray Affair." *Scientific American* 242: 122–131.

Koedinger, K. R., M. W. Alibali, and M. J. Nathan. 2008. "Trade-offs between Grounded and Abstract Representations: Evidence from Algebra Problem Solving." *Cognitive Science* 32: 366–397.

Koedinger, K. R., and J. R. Anderson. 1990. "Abstract Planning and Perceptual Chunks: Elements of Expertise in Geometry." *Cognitive Science* 14: 511–550.

Komatsu, L. K., and K. M. Galotti. 1986. "Children's Reasoning about Social, Physical, and Logical Regularities: A Look at Two Worlds." *Child Development* 57: 413–420.

Koslowski, B. 1996. *Theory and Evidence: The Development of Scientific Reasoning*. Cambridge, MA: MIT Press.

Koslowski, B., J. Marasia, M. Chelenza, and R. Dublin. 2008. "Information Becomes Evidence When an Explanation Can Incorporate It into a Causal Framework." *Cognitive Development* 23: 472–487.

Koslowski, B., L. Okagaki, C. Lorenz, and D. Umbach. 1989. "When Covariation Is Not Enough: The Role of Causal Mechanism, Sampling Method, and Sample Size in Causal Reasoning." *Child Development* 60: 1316–1327.

Kouchaki, M., I. H. Smith, and K. Savani. 2018. "Does Deciding among Morally Relevant Options Feel Like Making a Choice? How Morality Constrains People's Sense of Choice." *Journal of Personality and Social Psychology* 115: 788–804.

Kuhn, D. 1991. *The Skills of Argument*. Cambridge: Cambridge University Press.

Kuhn, D., E. Amsel, M. O'Loughlin, L. Schauble, B. Leadbeater, and W. Yotive. 1988. *The Development of Scientific Thinking Skills*. Cambridge, MA: Academic Press.

Kuhn, T. S. 1962. *The Structure of Scientific Revolutions*. Chicago: University of Chicago Press.

Kynast, J., E. M. Quinque, M., T.Polyakova, Luck, S. G. Riedel-Heller, S. Baron-Cohen, A. Hinz, A. V. Witte, J. Sacher, A. Villringer, and M. L. Schroeter. 2020. "Mindreading from the Eyes Declines with Aging: Evidence from 1,603 Subjects." *Frontiers in Aging Neuroscience* 12: 550416.

LaCour, M. J., and D. P. Green. 2014. "When Contact Changes Minds: An Experiment on Transmission of Support for Gay Equality." *Science* 346: 1366–1369.

Lancy, D. F. 2016. "Playing with Knives: The Socialization of Self-initiated Learners." *Child Development* 87: 654–665.

Lane, A. 2006. "Wonderful World: What Walt Disney Made." *New Yorker*, December 11. https://www.newyorker.com/magazine/2006/12/11/wonderful-world.

Lane, J. D., E. M. Evans, K. A. Brink, and H. M. Wellman. 2016. "Developing Concepts of Ordinary and Extraordinary Communication." *Developmental Psychology* 52: 19–30.

Lane, J. D., S. Ronfard, and E. El-Sherif. 2018. "The Influence of First-hand Testimony and Hearsay on Children's Belief in the Improbable." *Child Development* 89: 1133–1140.

Lane, J. D., S. Ronfard, S. P. Francioli, and P. L. Harris. 2016. "Children's Imagination and Belief: Prone to Flights of Fancy or Grounded in Reality?" *Cognition* 152: 127–140.

Lane, J. D., H. M. Wellman, and E. M. Evans. 2010. "Children's Understanding of Ordinary and Extraordinary Minds." *Child Development* 81: 1475–1489.

———. 2014. "Approaching an Understanding of Omniscience from the Preschool Years to Early Adulthood." *Developmental Psychology* 50: 2380–2392.

Lane, J. D., L. Zhu, E. M. Evans, and H. M. Wellman. 2016. "Developing Concepts of the Mind, Body, and Afterlife: Exploring the Roles of Narrative Context and Culture." *Journal of Cognition and Culture* 16: 50–82.

Lane, L. 2022. "DALL-E, Make Me Another Picasso, Please." *New Yorker*, July 4. https://www.newyorker.com/magazine/2022/07/11/dall-e-make-me-another-picasso-please.

Larsen, N. E., K. Lee, and P. A. Ganea. 2018. "Do Storybooks with Anthropomorphized Animal Characters Promote Prosocial Behaviors in Young Children?" *Developmental Science* 21: e12590.

Lascaux, A. 2020. "Of Kids and Unicorns: How Rational Is Children's Trust in Testimonial Knowledge?" *Cognitive Science* 44: e12819.

Latson, J. 2017. "How a Real Genetic Disorder Could Have Inspired Fairy Tales." *Time*, June 20. https://time.com/4823574/mythology-williams-syndrome/.

Lawson, R. 2006. "The Science of Cycology: Failures to Understand How Everyday Objects Work." *Memory and Cognition* 34: 1667–1675.

Leahy, B. P., and S. E. Carey. 2020. "The Acquisition of Modal Concepts." *Trends in Cognitive Sciences* 24: 65–78.

Leevers, H. J., and P. L. Harris. 1999. "Persisting Effects of Instruction on Young Children's Syllogistic Reasoning with Incongruent and Abstract Premises." *Thinking and Reasoning* 5: 145–173.

Legare, C. H., and M. Nielsen. 2015. "Imitation and Innovation: The Dual Engines of Cultural Learning." *Trends in Cognitive Sciences* 19: 688–699.

Lesage, K. A., and R. A. Richert. 2021. "Can God Do the Impossible? Anthropomorphism and Children's Certainty That God Can Make Impossible Things Possible." *Cognitive Development* 58: 101034.

Lesher, J. H. 1992. *Xenophanes of Colophon: Fragments*. Toronto: University of Toronto Press.

Levy, G. D., M. G. Taylor, and S. A. Gelman. 1995. "Traditional and Evaluative Aspects of Flexibility in Gender Roles, Social Conventions, Moral Rules, and Physical Laws." *Child Development* 66: 515–531.

Levy, N. 2017. "Religious Beliefs Are Factual Beliefs: Content Does Not Correlate with Context Sensitivity." *Cognition* 161: 109–116.

Lewis, D. 1978. "Truth in Fiction." *American Philosophical Quarterly* 15: 37–46.

Lewis, E. L., J. L. Stern, and M. C. Linn. 1993. "The Effect of Computer Simulations on Introductory Thermodynamics Understanding." *Educational Technology* 33: 45–58.

Lew-Levy, S., R. Reckin, N. Lavi, J. Cristóbal-Azkarate, and K. Ellis-Davies. 2017. "How Do Hunter-Gatherer Children Learn Subsistence Skills?" *Human Nature* 28: 367–394.

Lewry, C., K. Curtis, N. Vasilyeva, F. Xu, and T. L. Griffiths. 2021. "Intuitions about Magic Track the Development of Intuitive Physics." *Cognition* 214: 104762.

Li, H., K. Boguszewski, and A. S. Lillard. 2015. "Can That Really Happen? Children's Knowledge about the Reality Status of Fantastical Events in Television." *Journal of Experimental Child Psychology* 139: 99–114.

Li, H., T. Liu, J. Woolley, and P. Zhang. 2019. "Reality Status Judgments of Real and Fantastical Events in Children's Prefrontal Cortex: An fNIRS Study." *Frontiers in Human Neuroscience* 13: 444.

Liberman, N., O. Polack, B. Hameiri, and M. Blumenfeld. 2012. "Priming of Spatial Distance Enhances Children's Creative Performance." *Journal of Experimental Child Psychology* 111: 663–670.

Lillard, A. S. 2001. "Pretend Play as Twin Earth: A Social-Cognitive Analysis." *Developmental Review* 21: 495–531.

———. 2017. *Montessori: The Science behind the Genius*. New York: Oxford University Press.

Lillard, A. S., M. D. Lerner, E. J. Hopkins, R. A. Dore, E. D. Smith, and C. M. Palmquist. 2013. "The Impact of Pretend Play on Children's Development: A Review of the Evidence." *Psychological Bulletin* 139: 1–34.

Limon, M. 2001. "On Cognitive Conflict as an Instructional Strategy for Conceptual Change: A Critical Appraisal." *Learning and Instruction* 11: 357–380.

Lin, T. J., R. C. Anderson, J. E. Hummel, M. Jadallah, B. W. Miller, K. Nguyen-Jahiel, J. A. Morris, L-J. Kuo, I-H. Kim, X. Wu, and T. Dong. 2012. "Children's Use of Analogy during Collaborative Reasoning." *Child Development* 83: 1429–1443.

Lindstedt, J. K., and W. D. Gray. 2019. "Distinguishing Experts from Novices by the Mind's Hand and Mind's Eye." *Cognitive Psychology* 109: 1–25.

Lindstrom, B., S. Jangard, I. Selbing, and A. Olsson. 2018. "The Role of a 'Common Is Moral' Heuristic in the Stability and Change of Moral Norms." *Journal of Experimental Psychology: General* 147: 228–242.

Linsey, J. S., I. Tseng, K. Fu, J. Cagan, K. L. Wood, and C. Schunn. 2010. "A Study of Design Fixation, Its Mitigation and Perception in Engineering Design Faculty." *Journal of Mechanical Design* 132: 041003.

Liquin, E. G., and A. Gopnik. 2022. "Children Are More Exploratory and Learn More Than Adults in an Approach-Avoid Task." *Cognition* 218: 104940.

Liquin, E. G., and T. Lombrozo. 2020. "A Functional Approach to Explanation-Seeking Curiosity." *Cognitive Psychology* 119: 101276.

Liquin, E. G., S. E. Metz, and T. Lombrozo. 2020. "Science Demands Explanation, Religion Tolerates Mystery." *Cognition* 204: 104398.

Lockhart, K. L., B. Abrahams, and D. N. Osherson. 1977. "Children's Understanding of Uniformity in the Environment." *Child Development* 48: 1521–1531.

Lockhart, K. L., A. Chuey, S. Kerr, and F. C. Keil. 2019. "The Privileged Status of Knowing Mechanistic Information: An Early Epistemic Bias." *Child Development* 90: 1772–1788.

Loeb, A. 2021. *Extraterrestrial: The First Sign of Intelligent Life beyond Earth.* Boston: Mariner.

Logothetis, N. K., and J. Pauls. 1995. "Psychophysical and Physiological Evidence for Viewer-centered Object Representations in the Primate." *Cerebral Cortex* 5: 270–288.

Lombrozo, T. 2009. "The Role of Moral Commitments in Moral Judgment." *Cognitive Science* 33: 273–286.

Lopez-Mobilia, G., and J. D. Woolley. 2016. "Interactions between Knowledge and Testimony in Children's Reality-Status Judgments." *Journal of Cognition and Development* 17: 486–504.

Lucas, C. G., S. Bridgers, T. L. Griffiths, and A. Gopnik. 2014. "When Children Are Better (or at Least More Open-Minded) Learners Than Adults: Developmental Differences in Learning the Forms of Causal Relationships." *Cognition* 131: 284–299.

Luchins, A. S. 1942. "Mechanization in Problem Solving: The Effect of Einstellung." *Psychological Monographs* 54: i–95.

Luhrmann, T. M. 2020. *How God Becomes Real: Kindling the Presence of Invisible Others*. Princeton, NJ: Princeton University Press.

Lupfer, M. B., D. Tolliver, and M. Jackson. 1996. Explaining Life-altering Occurrences: A Test of the 'God-of-the-Gaps' Hypothesis." *Journal for the Scientific Study of Religion* 35: 379–391.

Lyons, D. E., A. G. Young, and F. C. Keil. 2007. "The Hidden Structure of Overimitation." *Proceedings of the National Academy of Sciences of the United States of America* 104: 19751–19756.

Mar, R. A., and K. Oatley. 2008. "The Function of Fiction Is the Abstraction and Simulation of Social Experience." *Perspectives on Psychological Science* 3: 173–192.

Mar, R. A., K. Oatley, J. Hirsh, J. Dela Paz, and J. B. Peterson. 2006. "Bookworms versus Nerds: Exposure to Fiction versus Non-fiction, Divergent Associations with Social Ability, and the Simulation of Fictional Social Worlds." *Journal of Research in Personality* 40: 694–712.

Mares, M. L., and E. E. Acosta. 2008. "Be Kind to Three-legged Dogs: Children's Literal Interpretations of TV's Moral Lessons." *Media Psychology* 11: 377–399.

Mares, M. L., and G. Sivakumar. 2014. "'Vámonos Means Go, but That's Made up for the Show': Reality Confusions and Learning from Educational TV." *Developmental Psychology* 50: 2498–2511.

Marsh, E. J., and L. K. Fazio. 2006. "Learning Errors from Fiction: Difficulties in Reducing Reliance on Fictional Stories." *Memory and Cognition* 34: 1140–1149.

Martin, J. W., M. Buon, and F. Cushman. 2021. "The Effect of Cognitive Load on Intent-based Moral Judgment." *Cognitive Science* 45: e12965.

Masson, M. E., D. N. Bub, and C. E. Lalonde. 2011. "Video-Game Training and Naïve Reasoning about Object Motion." *Applied Cognitive Psychology* 25: 166–173.

Mayer, R. E. 2004. "Should There Be a Three-Strikes Rule against Pure Discovery Learning?" *American Psychologist* 59: 14–19.

Mayor, A. 2011. *The First Fossil Hunters: Dinosaurs, Mammoths, and Myth in Greek and Roman Times*. Princeton, NJ: Princeton University Press.

McCaffrey, T. 2012. "Innovation Relies on the Obscure: A Key to Overcoming the Classic Problem of Functional Fixedness." *Psychological Science* 23: 215–218.

McCoy, J., and T. Ullman. 2019. "Judgments of Effort for Magical Violations of Intuitive Physics." *PLoS One* 14: e0217513.

McGuigan, N. 2013. "The Influence of Model Status on the Tendency of Young Children to Over-imitate." *Journal of Experimental Child Psychology* 116: 962–969.

McGuigan, N., J. Makinson, and A. Whiten. 2011. "From Over-imitation to Super-copying: Adults Imitate Causally Irrelevant Aspects of

Tool Use with Higher Fidelity Than Young Children." *British Journal of Psychology* 102: 1–18.

McGuigan, N., A. Whiten, E. Flynn, and V. Horner. 2007. "Imitation of Causally Necessary versus Unnecessary Tool Use by 3-and 5-Year-Old Children." *Cognitive Development* 22: 353–364.

McNamara, R. A., A. K. Willard, A. Norenzayan, and J. Henrich. 2019. "Weighing Outcome vs. Intent across Societies: How Cultural Models of Mind Shape Moral Reasoning." *Cognition* 182: 95–108.

Meder, B., C. M. Wu, E. Schulz, and A. Ruggeri. 2021. "Development of Directed and Random Exploration in Children." *Developmental Science* 24: e13095.

Medina, E. 2022. "Pastor Resigns after Incorrectly Performing Thousands of Baptisms." *New York Times*, February 14. https://www.nytimes.com/2022/02/14/us/catholic-priest-baptisms-phoenix.html.

Miller, C. S., J. F. Lehman, and K. R. Koedinger. 1999. "Goals and Learning in Microworlds." *Cognitive Science* 23: 305–336.

Millman, A. B., and C. L. Smith. 1997. "Darwin's Use of Analogical Reasoning in Theory Construction." *Metaphor and Symbol* 12: 159–187.

Mills, C. M. 2013. "Knowing When to Doubt: Developing a Critical Stance When Learning from Others." *Developmental Psychology* 49: 404–418.

Mills, C. M., and F. C. Keil. 2004. "Knowing the Limits of One's Understanding: The Development of an Awareness of an Illusion of Explanatory Depth." *Journal of Experimental Child Psychology* 87: 1–32.

Mills, C. M., K. R. Sands, S. P. Rowles, and I. L. Campbell. 2019. "'I Want to Know More!': Children Are Sensitive to Explanation Quality When Exploring New Information." *Cognitive Science* 43: e12706.

Mody, S., and S. Carey. 2016. "The Emergence of Reasoning by the Disjunctive Syllogism in Early Childhood." *Cognition* 154: 40–48.

Montessori, M. (1915) 1997. *The California Lectures of Maria Montessori: Collected Speeches and Writings.* Oxford: Clio Press.

Moran, J. M., L. L. Young, R. Saxe, S. M. Lee, D. O'Young, P. L. Mavros, and J. D. Gabrieli. 2011. "Impaired Theory of Mind for Moral Judgment in High-functioning Autism." *Proceedings of the National Academy of Sciences of the United States of America* 108: 2688–2692.

Morelli, G. A., B. Rogoff, and C. Angelillo. 2003. "Cultural Variation in Young Children's Access to Work or Involvement in Specialised Child-Focused Activities." *International Journal of Behavioral Development* 27: 264–274.

Moskal, B. M., and M. E. Magone. 2000. "Making Sense of What Students Know: Examining the Referents, Relationships and Modes Students Displayed in Response to a Decimal Task." *Educational Studies in Mathematics* 43: 313–335.

Moss, J., and R. Case. 1999. "Developing Children's Understanding of the Rational Numbers: A New Model and an Experimental Curriculum." *Journal for Research in Mathematics Education* 30: 122–147.

Muller, D. A. 2011. "Khan Academy and the Effectiveness of Science Videos." *Veritasium*, March 17. YouTube video, 08:03. https://www.youtube.com/watch?v=eVtCO84MDj8.

Muller, D. A., J. Bewes, M. D. Sharma, and P. Reimann. 2008. "Saying the Wrong Thing: Improving Learning with Multimedia by Including Misconceptions." *Journal of Computer Assisted Learning* 24: 144–155.

Murphy, F. C., G. Wilde, N. Ogden, P. J. Barnard, and A. J. Calder. 2009. "Assessing the Automaticity of Moral Processing: Efficient Coding of Moral Information during Narrative Comprehension." *Quarterly Journal of Experimental Psychology* 62: 41–49.

Nancekivell, S. E., and O. Friedman. 2017. "She Bought the Unicorn from the Pet Store: Six-to Seven-Year-Olds Are Strongly Inclined to Generate Natural Explanations." *Developmental Psychology* 53: 1079–1087.

National Science Board. 2018. *Science and Engineering Indicators*. Alexandria, VA: National Science Foundation.

Nettle, D. 2005. "The Wheel of Fire and the Mating Game: Explaining the Origins of Tragedy and Comedy." *Journal of Cultural and Evolutionary Psychology* 3: 39–56.

Newcombe, N. S., and T. F. Shipley. 2015. "Thinking about Spatial Thinking: New Typology, New Assessments." In *Studying Visual and Spatial Reasoning for Design Creativity*, edited by J. S. Gero, pp. 179–192. Dordrecht, the Netherlands: Springer.

Nguyen, C. T. 2017. "Philosophy of Games." *Philosophy Compass* 12: e12426.

Nichols, S. 2006. "Imaginative Blocks and Impossibility: An Essay in Modal Psychology." In *The Architecture of the Imagination*, edited by S. Nichols, 237–255. New York: Oxford University Press.

Nichols, S., and S. Stich. 2000. "A Cognitive Theory of Pretense." *Cognition* 74: 115–147.

Nieder, A. 2021. "The Evolutionary History of Brains for Numbers." *Trends in Cognitive Sciences* 25: 608–621.

Nielsen, M. 2013. "Young Children's Imitative and Innovative Behaviour on the Floating Object Task." *Infant and Child Development* 22: 44–52.

Nielsen, M., and C. Blank. 2011. "Imitation in Young Children: When Who Gets Copied Is More Important Than What Gets Copied." *Developmental Psychology* 47: 1050–1053.

Nielsen, M., K. Tomaselli, I. Mushin, and A. Whiten. 2014. "Exploring Tool Innovation: A Comparison of Western and Bushman Children." *Journal of Experimental Child Psychology* 126: 384–394.

Nolan-Reyes, C., M. A. Callanan, and K. A. Haigh. 2016. "Practicing Possibilities: Parents' Explanations of Unusual Events and Children's Possibility Thinking." *Journal of Cognition and Development* 17: 378–395.

Norenzayan, A. 2013. *Big Gods: How Religion Transformed Cooperation and Conflict*. Princeton, NJ: Princeton University Press.

Norenzayan, A., S. Atran, J. Faulkner, and M. Schaller. 2006. "Memory and Mystery: The Cultural Selection of Minimally Counterintuitive Narratives." *Cognitive Science* 30: 531–553.

Novick, L. R., and K. J. Holyoak. 1991. "Mathematical Problem Solving by Analogy." *Journal of Experimental Psychology: Learning, Memory, and Cognition* 17: 398–415.

Nyhof, M. A., and C. N. Johnson. 2017. "Is God Just a Big Person? Children's Conceptions of God across Cultures and Religious Traditions." *British Journal of Developmental Psychology* 35: 60–75.

Nyhout, A., and P. A. Ganea. 2021. "Scientific Reasoning and Counterfactual Reasoning in Development." *Advances in Child Development and Behavior* 61: 223–253.

Obersteiner, A., W. Van Dooren, J. Van Hoof, and L. Verschaffel. 2013. "The Natural Number Bias and Magnitude Representation in Fraction Comparison by Expert Mathematicians." *Learning and Instruction* 28: 64–72.

O'Keane, V. 2021. *A Sense of Self: Memory, the Brain, and Who We Are*. New York: W.W. Norton.

Onishi, K. H., R. Baillargeon, and A. M. Leslie. 2007. "15-Month-Old Infants Detect Violations in Pretend Scenarios." *Acta Psychologica* 124: 106–128.

Oost, S. I. 1975. "Thucydides and the Irrational: Sundry Passages." *Classical Philology* 70: 186–196.

Orben, A. 2020. "The Sisyphean Cycle of Technology Panics." *Perspectives on Psychological Science* 15: 1143–1157.

Oreskes, N. 1999. *The Rejection of Continental Drift: Theory and Method in American Earth Science*. New York: Oxford University Press.

Ormand, C. J., C. Manduca, T. F. Shipley, B. Tikoff, C. L. Harwood, K. Atit, and A. P. Boone. 2014. "Evaluating Geoscience Students' Spatial Thinking Skills in a Multi-institutional Classroom Study." *Journal of Geoscience Education* 62: 146–154.

Orozco-Giraldo, C., and P. L. Harris. 2019. "Turning Water into Wine: Young Children's Conception of the Impossible." *Journal of Cognition and Culture* 19: 219–243.

Ozturk, O., and A. Papafragou. 2015. "The Acquisition of Epistemic Modality: From Semantic Meaning to Pragmatic Interpretation." *Language Learning and Development* 11: 191–214.

Packard, M. 2013. *Little Raccoon Learns to Share*. New York: Sterling Children's Books.

Panero, M. E., D. S. Weisberg, J. Black, T. R. Goldstein, J. L. Barnes, H. Brownell, and E. Winner. 2016. "Does Reading a Single Passage of Literary Fiction Really Improve Theory of Mind? An Attempt at Replication." *Journal of Personality and Social Psychology* 111: e46–e54.

Paulus, M., A. Nöth, and M. Wörle. 2018. "Preschoolers' Resource Allocations Align with Their Normative Judgments." *Journal of Experimental Child Psychology* 175: 117–126.

Payir, A., L. Heiphetz, P. L. Harris, and K. H. Corriveau. 2022. "What Could Have Been Done? Counterfactual Alternatives to Negative Outcomes Generated by Religious and Secular Children." *Developmental Psychology* 58: 376–391.

Payir, A., N. McLoughlin, Y. K. Cui, T. Davoodi, J. M. Clegg, P. L. Harris, and K. H. Corriveau. 2021. "Children's Ideas about What Can Really Happen: The Impact of Age and Religious Background." *Cognitive Science* 45: e13054.

Pennington, N., and R. Hastie. 1986. "Evidence Evaluation in Complex Decision Making." *Journal of Personality and Social Psychology* 51: 242–258.

Peoples, H. C., P. Duda, and F. W. Marlowe. 2016. "Hunter-Gatherers and the Origins of Religion." *Human Nature* 27: 261–282.

Periss, V., C. H. Blasi, and D. F. Bjorklund. 2012. "Cognitive 'Babyness': Developmental Differences in the Power of Young Children's Supernatural Thinking to Influence Positive and Negative Affect." *Developmental Psychology* 48: 1203–1214.

Peterson, M. 2017. *An Introduction to Decision Theory*. Cambridge: Cambridge University Press.

Phelps, K. E., and J. D. Woolley. 1994. "The Form and Function of Young Children's Magical Beliefs." *Developmental Psychology* 30: 385–394.

Phillips, J., and F. Cushman. 2017. "Morality Constrains the Default Representation of What Is Possible." *Proceedings of the National Academy of Sciences of the United States of America* 114: 4649–4654.

Phillips, J., and J. Knobe. 2018. "The Psychological Representation of Modality." *Mind and Language* 33: 65–94.

Phillips, J., A. Morris, and F. Cushman. 2019. "How We Know What Not to Think." *Trends in Cognitive Sciences* 23: 1026–1040.

Piaget, J. (1932) 1965. *The Moral Judgment of the Child*. New York: Free Press.

Pica, P., C. Lemer, V. Izard, and S. Dehaene. 2004. "Exact and Approximate Arithmetic in an Amazonian Indigene Group." *Science* 306: 499–503.

Prabhakar, J., and S. Ghetti. 2020. "Connecting the Dots between Past and Future: Constraints in Episodic Future Thinking in Early Childhood." *Child Development* 91: e315–e330.

Prather, R. W., and M. W. Alibali. 2008. "Understanding and Using Principles of Arithmetic: Operations Involving Negative Numbers." *Cognitive Science* 32: 445–457.

Prentice, N. M., and D. A. Gordon. 1987. "Santa Claus and the Tooth Fairy for the Jewish Child and Parent." *Journal of Genetic Psychology* 148: 139–151.

Prentice, N. M., M. Manosevitz, and L. Hubbs. 1978. "Imaginary Figures of Early Childhood: Santa Claus, Easter Bunny, and the Tooth Fairy." *American Journal of Orthopsychiatry* 48: 618–628.

Purzycki, B. G., D. N. Finkel, J. Shaver, N. Wales, A. B. Cohen, and R. Sosis. 2012. "What Does God Know? Supernatural Agents' Access to Socially Strategic and Non-strategic Information." *Cognitive Science* 36: 846–869.

Rakoczy, H. 2008. "Taking Fiction Seriously: Young Children Understand the Normative Structure of Joint Pretense Games." *Developmental Psychology* 44: 1195–1201.

Rakoczy, H., M. Kaufmann, and K. Lohse. 2016. "Young Children Understand the Normative Force of Standards of Equal Resource Distribution." *Journal of Experimental Child Psychology* 150: 396–403.

Rakoczy, H., F. Warneken, and M. Tomasello. 2008. "The Sources of Normativity: Young Children's Awareness of the Normative Structure of Games." *Developmental Psychology* 44: 875–881.

Ranney, M. A., and D. Clark. 2016. "Climate Change Conceptual Change: Scientific Information Can Transform Attitudes." *Topics in Cognitive Science* 8: 49–75.

Rawlings, B., and C. H. Legare. 2021. "Toddlers, Tools, and Tech: The Cognitive Ontogenesis of Innovation." *Trends in Cognitive Sciences* 25: 81–92.

Redshaw, J., T. Leamy, P. Pincus, and T. Suddendorf. 2018. "Young Children's Capacity to Imagine and Prepare for Certain and Uncertain Future Outcomes." *PloS One* 13: e0202606.

Redshaw, J., and T. Suddendorf. 2016. "Children's and Apes' Preparatory Responses to Two Mutually Exclusive Possibilities." *Current Biology* 26: 1758–1762.

Reindl, E., S. R. Beck, I. A. Apperly, and C. Tennie. 2016. "Young Children Spontaneously Invent Wild Great Apes' Tool-Use Behaviours." *Proceedings of the Royal Society B: Biological Sciences* 283: 20152402.

Renken, M. D., and N. Nunez. 2010. "Evidence for Improved Conclusion Accuracy after Reading about Rather Than Conducting a Belief-Inconsistent Simple Physics Experiment." *Applied Cognitive Psychology* 24: 792–811.

———. 2013. "Computer Simulations and Clear Observations Do Not Guarantee Conceptual Understanding." *Learning and Instruction* 23: 10–23.

Reuter, T., and M. Leuchter. 2021. "Children's Concepts of Gears and Their Promotion through Play." *Journal of Research in Science Teaching* 58: 69–94.

Richards, C. A., and J. A. Sanderson. 1999. "The Role of Imagination in Facilitating Deductive Reasoning in 2-, 3-and 4-Year-Olds." *Cognition* 72: B1–B9.

Richert, R. A., N. J. Shaman, A. R. Saide, and K. A. Lesage. 2016. "Folding Your Hands Helps God Hear You: Prayer and Anthropomorphism in Parents and Children." *Research in the Social Scientific Study of Religion* 27: 140–157.

Richert, R. A., A. B. Shawber, R. E. Hoffman, and M. Taylor. 2009. "Learning from Fantasy and Real Characters in Preschool and Kindergarten." *Journal of Cognition and Development* 10: 41–66.

Richert, R. A., and E. I. Smith. 2011. "Preschoolers' Quarantining of Fantasy Stories." *Child Development* 82: 1106–1119.

Ridley, M. 2020. *How Innovation Works: And Why It Flourishes in Freedom*. New York: Harper Collins.

Riggs, A. E., and C. W. Kalish. 2016. "Children's Evaluations of Rule Violators." *Cognitive Development* 40: 132–143.

Rizzo, M. T., L. Elenbaas, S. Cooley, and M. Killen. 2016. "Children's Recognition of Fairness and Others' Welfare in a Resource Allocation Task: Age Related Changes." *Developmental Psychology* 52: 1307–1317.

Roberts, S. O., S. A. Gelman, and A. K. Ho. 2017. "So It Is, So It Shall Be: Group Regularities License Children's Prescriptive Judgments." *Cognitive Science* 41: 576–600.

Roberts, S. O., C. Guo, A. K. Ho, and S. A. Gelman. 2018. "Children's Descriptive-to-Prescriptive Tendency Replicates (and Varies) Cross-culturally: Evidence from China." *Journal of Experimental Child Psychology* 165: 148–160.

Roberts, S. O., A. K. Ho, and S. A. Gelman. 2017. "Group Presence, Category Labels, and Generic Statements Influence Children to Treat Descriptive Group Regularities as Prescriptive." *Journal of Experimental Child Psychology* 158: 19–31.

Roberts, S. O., and R. I. Horii. 2019. "Thinking Fast and Slow: Children Are Less Negative toward Non-conformity When They Reflect before Responding." *Journal of Cognition and Development* 20: 790–799.

Robinson, K. 2007. "Do Schools Kill Creativity?" *TED*, January 7. YouTube video, 20:03. https://www.youtube.com/watch?v=iG9CE55wbtY.

Roby, A. C., and E. Kidd. 2008. "The Referential Communication Skills of Children with Imaginary Companions." *Developmental Science* 11: 531–540.

Ronfard, S., S. Brown, E. Doncaster, and D. Kelemen. 2021. "Inhibiting Intuition: Scaffolding Children's Theory Construction about Species Evolution in the Face of Competing Explanations." *Cognition* 211: 104635.

Root-Bernstein, M. M. 2013. "The Creation of Imaginary Worlds." In *The Oxford Handbook of the Development of Imagination*, edited by M. Taylor, 417–437. Oxford: Oxford University Press.

Rosengren, K. S., and A. K. Hickling. 1994. "Seeing Is Believing: Children's Explanations of Commonplace, Magical, and Extraordinary Transformations." *Child Development* 65: 1605–1626.

Ross, L. D., Y. Lelkes, and A. G. Russell. 2012. "How Christians Reconcile Their Personal Political Views and the Teachings of Their Faith: Projection as a Means of Dissonance Reduction." *Proceedings of the National Academy of Sciences of the United States of America* 109: 3616–3622.

Rotman, D. 2020. "We're Not Prepared for the End of Moore's Law." *MIT Technology Review*, February 24. https://www.technologyreview.com/2020/02/24/905789/were-not-prepared-for-the-end-of-moores-law/.

Rottman, B. M., D. Gentner, and M. B. Goldwater. 2012. "Causal Systems Categories: Differences in Novice and Expert Categorization of Causal Phenomena." *Cognitive Science* 36: 919–932.

Rottman, J., L. Young, and D. Kelemen. 2017. "The Impact of Testimony on Children's Moralization of Novel Actions." *Emotion* 17: 811–827.

Rottman, J., V. Zizik, K. Minard, L. Young, P. R. Blake, and D. Kelemen. 2020. "The Moral, or the Story? Changing Children's Distributive Justice Preferences through Social Communication." *Cognition* 205: 104441.

Rozenblit, L., and F. Keil. 2002. "The Misunderstood Limits of Folk Science: An Illusion of Explanatory Depth." *Cognitive Science* 26: 521–562.

Ryan, M. J. 2018. *A Taste for the Beautiful: The Evolution of Attraction*. Princeton, NJ: Princeton University Press.

Saide, A., and Richert, R. 2022. "Correspondence in Parents' and Children's Concepts of God: Investigating the Role of Parental Values, Religious Practices and Executive Functioning." *British Journal of Developmental Psychology* 40: 422–437.

Said-Metwaly, S., B. Fernández-Castilla, E. Kyndt, W. Van den Noortgate, and B. Barbot. 2021. "Does the Fourth-Grade Slump in Creativity Actually Exist? A Meta-analysis of the Development of Divergent Thinking in School-Age Children and Adolescents." *Educational Psychology Review* 33: 275–298.

Sarnecka, B. W., and S. Carey. 2008. "How Counting Represents Number: What Children Must Learn and When They Learn It." *Cognition* 108: 662–674.

Sarnecka, B. W., and M. D. Lee. 2009. "Levels of Number Knowledge during Early Childhood." *Journal of Experimental Child Psychology* 103: 325–337.

Schalk, L., H. Saalbach, and E. Stern. 2016. "Approaches to Foster Transfer of Formal Principles: Which Route to Take?" *PLoS One* 11: e0148787.

Schauble, L. 1996. "The Development of Scientific Reasoning in Knowledge-Rich Contexts." *Developmental Psychology* 32: 102–119.

Schjoedt, U., H. Stødkilde-Jørgensen, A. W. Geertz, and A. Roepstorff. 2009. "Highly Religious Participants Recruit Areas of Social Cognition in Personal Prayer." *Social Cognitive and Affective Neuroscience* 4: 199–207.

Schmidt, M. F., L. P. Butler, J. Heinz, and M. Tomasello. 2016. "Young Children See a Single Action and Infer a Social Norm: Promiscuous Normativity in 3-Year-Olds." *Psychological Science* 27: 1360–1370.

Schmidt, M. F., H. Rakoczy, T. Mietzsch, and M. Tomasello. 2016. "Young Children Understand the Role of Agreement in Establishing Arbitrary Norms—but Unanimity Is Key." *Child Development* 87: 612–626.

Schmidt, M. F., H. Rakoczy, and M. Tomasello. 2012. "Young Children Enforce Social Norms Selectively Depending on the Violator's Group Affiliation." *Cognition* 124: 325–333.

Schmidt, M. F., M. Svetlova, J. Johe, and M. Tomasello. 2016. "Children's Developing Understanding of Legitimate Reasons for Allocating Resources Unequally." *Cognitive Development* 37: 42–52.

Schulz, L. E., E. B. Bonawitz, and T. L. Griffiths. 2007. "Can Being Scared Cause Tummy Aches? Naive Theories, Ambiguous Evidence, and Preschoolers' Causal Inferences." *Developmental Psychology* 43: 1124–1139.

Schulz, L. E., N. D. Goodman, J. B. Tenenbaum, and A. C. Jenkins. 2008. "Going beyond the Evidence: Abstract Laws and Preschoolers' Responses to Anomalous Data." *Cognition* 109: 211–223.

Schumpeter, J. A. 1942. *Capitalism, Socialism and Democracy*. Manhattan: Harper & Brothers.

Sfard, A. 1991. "On the Dual Nature of Mathematical Conceptions: Reflections on Processes and Objects as Different Sides of the Same Coin." *Educational Studies in Mathematics* 22: 1–36.

Shaw, A., and K. R. Olson. 2012. "Children Discard a Resource to Avoid Inequity." *Journal of Experimental Psychology: General* 141: 382–395.

Shermer, M. 2010. "Living in Denial: When a Sceptic Isn't a Sceptic." *New Scientist*, May 12. https://www.newscientist.com/article/mg20627606-000-living-in-denial-when-a-sceptic-isnt-a-sceptic/.

Shtulman, A. 2006. "Qualitative Differences between Naïve and Scientific Theories of Evolution." *Cognitive Psychology* 52: 170–194.

———. 2008. "Variation in the Anthropomorphization of Supernatural Beings and Its Implications for Cognitive Theories of Religion." *Journal of Experimental Psychology: Learning, Memory, and Cognition* 34: 1123–1138.

———. 2009. "The Development of Possibility Judgment within and across Domains." *Cognitive Development* 24: 293–309.

———. 2013. "Epistemic Similarities between Students' Scientific and Supernatural Beliefs." *Journal of Educational Psychology* 105: 199–212.

———. 2017. *Scienceblind: Why Our Intuitive Theories about the World Are So Often Wrong*. New York: Basic Books.

Shtulman, A., and P. Calabi. 2012. "Cognitive Constraints on the Understanding and Acceptance of Evolution." In *Evolution Challenges: Integrating Research and Practice in Teaching and Learning about Evolution*, edited by K. S. Rosengren, S. Brem, E. M. Evans, and G. Sinatra, 47–65. Oxford: Oxford University Press.

Shtulman, A., and S. Carey. 2007. "Improbable or Impossible? How Children Reason about the Possibility of Extraordinary Events." *Child Development* 78: 1015–1032.

Shtulman, A., R. Foushee, D. Barner, Y. Dunham, and M. Srinivasan. 2019. "When Allah Meets Ganesha: Developing Supernatural Concepts in a Religiously Diverse Society." *Cognitive Development* 52: 100806.

Shtulman, A., and M. Lindeman. 2016. "Attributes of God: Conceptual Foundations of a Foundational Belief." *Cognitive Science* 40: 635–670.

Shtulman, A., and C. Morgan. 2017. "The Explanatory Structure of Unexplainable Events: Causal Constraints on Magical Reasoning." *Psychonomic Bulletin and Review* 24: 1573–1585.

Shtulman, A., C. Neal, and G. Lindquist. 2016. "Children's Ability to Learn Evolutionary Explanations for Biological Adaptation." *Early Education and Development* 27: 1222–1236.
Shtulman, A., and J. Phillips. 2018. "Differentiating 'Could' from 'Should': Developmental Changes in Modal Cognition." *Journal of Experimental Child Psychology* 165: 161–182.
Shtulman, A., and M. Rattner. 2018. "Theories of God: Explanatory Coherence in Religious Cognition." *PLoS One* 13: e0209758.
Shtulman, A., I. Share, R. Silber-Marker, and A. R. Landrum. 2020. "OMG GMO! Parent-Child Conversations about Genetically Modified Foods." *Cognitive Development* 55: 100895.
Shtulman, A., and L. Tong. 2013. "Cognitive Parallels between Moral Judgment and Modal Judgment." *Psychonomic Bulletin and Review* 20: 1327–1335.
Shtulman, A., and R. I. Yoo. 2015. "Children's Understanding of Physical Possibility Constrains Their Belief in Santa Claus." *Cognitive Development* 34: 51–62.
Shusterman, A., S. Ah Lee, and E. S. Spelke. 2008. "Young Children's Spontaneous Use of Geometry in Maps." *Developmental Science* 11: F1–F7.
Siegel, M. H., R. W. Magid, M. Pelz, J. B. Tenenbaum, and L. E. Schulz. 2021. "Children's Exploratory Play Tracks the Discriminability of Hypotheses." *Nature Communications* 12: 1–9.
Siegler, R. S., G. J. Duncan, P. E. Davis-Kean, K. Duckworth, A. Claessens, M. Engel, M. I. Susperreguy, and M. Chen. 2012. "Early Predictors of High School Mathematics Achievement." *Psychological Science* 23: 691–697.
Silva Luna, D., and J. M. Bering. 2021. "The Construction of Awe in Science Communication." *Public Understanding of Science* 30: 2–15.
Singer, D. G., J. L. Singer, H. D'Agnostino, and R. DeLong. 2009. "Children's Pastimes and Play in Sixteen Nations: Is Free-Play Declining?" *American Journal of Play* 1: 283–312.
Singer, J. L., and D. G. Singer. 2013. "Historical Overview of Research on Imagination in Children." In *The Oxford Handbook of the Development of Imagination*, edited by M. Taylor, 11–27. Oxford: Oxford University Press.
Singh, M. 2021. "The Sympathetic Plot, Its Psychological Origins, and Implications for the Evolution of Fiction." *Emotion Review* 13: 183–198.

Slone, J. 2004. *Theological Incorrectness: Why Religious People Believe What They Shouldn't*. Oxford: Oxford University Press.

Slotta, J. D., and M. T. Chi. 2006. "Helping Students Understand Challenging Topics in Science through Ontology Training." *Cognition and Instruction* 24: 261–289.

Slusser, E. B., and B. W. Sarnecka. 2011. "Find the Picture of Eight Turtles: A Link between Children's Counting and Their Knowledge of Number Word Semantics." *Journal of Experimental Child Psychology* 110: 38–51.

Smith, C. L., G. E. Solomon, and S. Carey. 2005. "Never Getting to Zero: Elementary School Students' Understanding of the Infinite Divisibility of Number and Matter." *Cognitive Psychology* 51: 101–140.

Smith, S. M., and S. E. Blankenship. 1991. "Incubation and the Persistence of Fixation in Problem Solving." *American Journal of Psychology* 104: 61–87.

Sobel, D. M. 2004. "Exploring the Coherence of Young Children's Explanatory Abilities: Evidence from Generating Counterfactuals." *British Journal of Developmental Psychology* 22: 37–58.

Sobel, D. M., and D. S. Weisberg. 2014. "Tell Me a Story: How Children's Developing Domain Knowledge Affects Their Story Construction." *Journal of Cognition and Development* 15: 465–478.

Spaepen, E., M. Coppola, E. S. Spelke, S. E. Carey, and S. Goldin-Meadow. 2011. "Number without a Language Model." *Proceedings of the National Academy of Sciences of the United States of America* 108: 3163–3168.

Spelke, E. S. 1994. "Initial Knowledge: Six Suggestions." *Cognition* 50: 431–445.

Spelke, E. S., S. A. Lee, and V. Izard. 2010. "Beyond Core Knowledge: Natural Geometry." *Cognitive Science* 34: 863–884.

Stafylidou, S., and S. Vosniadou. 2004. "The Development of Students' Understanding of the Numerical Value of Fractions." *Learning and Instruction* 14: 503–518.

Stahl, A. E., and L. Feigenson. 2015. "Observing the Unexpected Enhances Infants' Learning and Exploration." *Science* 348: 91–94.

Stanley, M. L., A. M. Dougherty, B. W. Yang, P. Henne, and F. De Brigard. 2018. "Reasons Probably Won't Change Your Mind: The Role of Reasons in Revising Moral Decisions." *Journal of Experimental Psychology: General* 147: 962–987.

Starmans, C., M. Sheskin, and P. Bloom. 2017. "Why People Prefer Unequal Societies." *Nature Human Behaviour* 1: 1–7.

Stiller, J., D. Nettle, and R. I. Dunbar. 2003. "The Small World of Shakespeare's Plays." *Human Nature* 14: 397–408.

Strogatz, S. 2013. *The Joy of X: A Guided Tour of Math, from One to Infinity.* Boston: Mariner.

Strohminger, N., and V. Kumar. 2018. *The Moral Psychology of Disgust.* Lanham, MD: Rowman & Littlefield.

Subbotsky, E. 1994. "Early Rationality and Magical Thinking in Preschoolers: Space and Time." *British Journal of Developmental Psychology* 12: 97–108.

Subiaul, F., E. Krajkowski, E. E. Price, and A. Etz. 2015. "Imitation by Combination: Preschool Age Children Evidence Summative Imitation in a Novel Problem-Solving Task." *Frontiers in Psychology* 6: 1410.

Subiaul, F., and M. A. Stanton. 2020. "Intuitive Invention by Summative Imitation in Children and Adults." *Cognition* 202: 104320.

Suddendorf, T. 2010. "Linking Yesterday and Tomorrow: Preschoolers' Ability to Report Temporally Displaced Events." *British Journal of Developmental Psychology* 28: 491–498.

Suddendorf, T., and A. Dong. 2013. "On the Evolution of Imagination and Design." In *The Oxford Handbook of the Development of Imagination*, edited by M. Taylor, 453–467. Oxford: Oxford University Press.

Sutherland, S. L., and O. Friedman. 2013. "Just Pretending Can Be Really Learning: Children Use Pretend Play as a Source for Acquiring Generic Knowledge." *Developmental Psychology* 49: 1660–1668.

Taggart, J., I. Becker, J. Rauen, H. Al Kallas, and A. S. Lillard. 2020. "What Shall We Do: Pretend or Real? Preschoolers' Choices and Parents' Perceptions." *Journal of Cognition and Development* 21: 261–281.

Taggart, J., M. J. Heise, and A. S. Lillard. 2018. "The Real Thing: Preschoolers Prefer Actual Activities to Pretend Ones." *Developmental Science* 21: e12582.

Taylor, M., and S. M. Carlson. 1997. "The Relation between Individual Differences in Fantasy and Theory of Mind." *Child Development* 68: 436–455.

Taylor, M., B. S. Cartwright, and S. M. Carlson. 1993. "A Developmental Investigation of Children's Imaginary Companions." *Developmental Psychology* 29: 276–285.

Taylor, M., C. M. Mottweiler, N. R. Aguiar, E. R. Naylor, and J. G. Levernier. 2020. "Paracosms: The Imaginary Worlds of Middle Childhood." *Child Development* 91: e164–e178.

Tennie, C., J. Call, and M. Tomasello. 2009. "Ratcheting Up the Ratchet: On the Evolution of Cumulative Culture." *Philosophical Transactions of the Royal Society B: Biological Sciences* 364: 2405–2415.

Tennie, C., V. Walter, A. Gampe, M. Carpenter, and M. Tomasello. 2014. "Limitations to the Cultural Ratchet Effect in Young Children." *Journal of Experimental Child Psychology* 126: 152–160.

Theobald, M., and G. Brod. 2021. "Tackling Scientific Misconceptions: The Element of Surprise." *Child Development* 92: 2128–2141.

Thompson, B. N., and T. R. Goldstein. 2020. "Children Learn from Both Embodied and Passive Pretense: A Replication and Extension." *Child Development* 91: 1364–1374.

Thompson, T. J., H. E. Winn, and P. J. Perkins. 1979. "Mysticete Sounds." In *Behavior of Marine Animals*, edited by H. E. Winn and B. L. Olla, 403–431. New York: Springer.

Thorburn, R., C. K. Bowman-Smith, and O. Friedman. 2020. "Likely Stories: Young Children Favor Typical over Atypical Story Events." *Cognitive Development* 56: 100950.

Tian, J., D. W. Braithwaite, and R. S. Siegler. 2021. "Distributions of Textbook Problems Predict Student Learning: Data from Decimal Arithmetic." *Journal of Educational Psychology* 113: 516–529.

Torbeyns, J., M. Schneider, Z. Xin, and R. S. Siegler. 2015. "Bridging the Gap: Fraction Understanding Is Central to Mathematics Achievement in Students from Three Different Continents." *Learning and Instruction* 37: 5–13.

Travers, P. L. 1934. *Mary Poppins*. New York: HarperCollins.

Trionfi, G., and E. Reese. 2009. "A Good Story: Children with Imaginary Companions Create Richer Narratives." *Child Development* 80: 1301–1313.

Twain, M. 1897. *Following the Equator*. Hartford, CT: American Publishing Company.

Tworek, C. M., and A. Cimpian. 2016. "Why Do People Tend to Infer 'Ought' from 'Is'? The Role of Biases in Explanation." *Psychological Science* 27: 1109–1122.

Tzelgov, J., D. Ganor-Stern, and K. Maymon-Schreiber. 2009. "The Representation of Negative Numbers: Exploring the Effects of

Mode of Processing and Notation." *Quarterly Journal of Experimental Psychology* 62: 605–624.

Vaden, V. C., and J. D. Woolley. 2011. "Does God Make It Real? Children's Belief in Religious Stories from the Judeo-Christian Tradition." *Child Development* 82: 1120–1135.

Vaesen, K. 2012. "The Cognitive Bases of Human Tool Use." *Behavioral and Brain Sciences* 35: 203–218.

Valdesolo, P., A. Shtulman, and A. S. Baron. 2017. "Science Is Awe-some: The Emotional Antecedents of Science Learning." *Emotion Review* 9: 1–7.

Vamvakoussi, X., W. Van Dooren, and L. Verschaffel. 2012. "Naturally Biased? In Search for Reaction Time Evidence for a Natural Number Bias in Adults." *Journal of Mathematical Behavior* 31: 344–355.

Van de Vondervoort, J. W., and O. Friedman. 2017. "Young Children Protest and Correct Pretense That Contradicts Their General Knowledge." *Cognitive Development* 43: 182–189.

Van Leeuwen, N. 2014. "Religious Credence Is Not Factual Belief." *Cognition* 133: 698–715.

Van Leeuwen, N., K. Weisman, and T. M. Luhrmann. 2021. "To Believe Is Not to Think: A Cross-cultural Finding." *Open Mind* 5: 91–99.

Van Reet, J., A. M. Pinkham, and A. S. Lillard. 2015. "The Effect of Realistic Contexts on Ontological Judgments of Novel Entities." *Cognitive Development* 34: 88–98.

Van Schijndel, T. J., I. Visser, B. M. van Bers, and M. E. Raijmakers. 2015. "Preschoolers Perform More Informative Experiments after Observing Theory-Violating Evidence." *Journal of Experimental Child Psychology* 131: 104–119.

Varma, S., and S. R. Karl. 2013. "Understanding Decimal Proportions: Discrete Representations, Parallel Access, and Privileged Processing of Zero." *Cognitive Psychology* 66: 283–301.

Vivanti, G., D. R. Hocking, P. Fanning, and C. Dissanayake. 2017. "The Social Nature of Overimitation: Insights from Autism and Williams Syndrome." *Cognition* 161: 10–18.

Wakslak, C. J. 2012. "The Where and When of Likely and Unlikely Events." *Organizational Behavior and Human Decision Processes* 117: 150–157.

Walden, T., G. Kim, C. McCoy, and J. Karrass. 2007. "Do You Believe in Magic? Infants' Social Looking during Violations of Expectations." *Developmental Science* 10: 654–663.

Walker, C. M., E. Bonawitz, and T. Lombrozo. 2017. "Effects of Explaining on Children's Preference for Simpler Hypotheses." *Psychonomic Bulletin and Review* 24: 1538–1547.

Walker, C. M., A. Gopnik, and P. A. Ganea. 2015. "Learning to Learn from Stories: Children's Developing Sensitivity to the Causal Structure of Fictional Worlds. *Child Development* 86: 310–318.

Walker, C. M., and T. Lombrozo. 2017. "Explaining the Moral of the Story." *Cognition* 167: 266–281.

Walker-Andrews, A. S., and P. L. Harris. 1993. "Young Children's Comprehension of Pretend Causal Sequences." *Developmental Psychology* 29: 915–921.

Ward, T. B. 1994. "Structured Imagination: The Role of Category Structure in Exemplar Generation." *Cognitive Psychology* 27: 1–40.

Watts, J., S. Passmore, J. C. Jackson, C. Rzymski, and R. I. Dunbar. 2020. "Text Analysis Shows Conceptual Overlap as Well as Domain-specific Differences in Christian and Secular Worldviews." *Cognition* 201: 104290.

Weber, R. J., and S. Dixon. 1989. "Invention and Gain Analysis." *Cognitive Psychology* 21: 283–302.

Weisberg, D. S., and J. Goodstein. 2009. "What Belongs in a Fictional World." *Journal of Cognition and Culture* 9: 69–78.

Weisberg, D. S., and A. Gopnik. 2013. "Pretense, Counterfactuals, and Bayesian Causal Models: Why What Is Not Real Really Matters." *Cognitive Science* 37: 1368–1381.

Weisberg, D. S., K. Hirsh-Pasek, R. M. Golinkoff, A. K. Kittredge, and D. Klahr. 2016. "Guided Play: Principles and Practices." *Current Directions in Psychological Science* 25: 177–182.

Weisberg, D. S., A. R. Landrum, S. E. Metz, and M. Weisberg. 2018. "No Missing Link: Knowledge Predicts Acceptance of Evolution in the United States." *BioScience* 68: 212–222.

Weisberg, D. S., D. M. Sobel, J. Goodstein, and P. Bloom. 2013. "Young Children Are Reality-Prone When Thinking about Stories." *Journal of Cognition and Culture* 13: 383–407.

Weisberg, R. W. 1999. "Creativity and Knowledge: A Challenge to Theories. In *Handbook of Creativity*, edited by R. J. Sternberg, 226–250. Cambridge: Cambridge University Press.

Weisman, K., and E. M. Markman. 2017. "Theory-based Explanation as Intervention." *Psychonomic Bulletin and Review* 24: 1555–1562.

Wenger, J. L. 2001. "Children's Theories of God: Explanations for Difficult-to-Explain Phenomena." *Journal of Genetic Psychology* 162: 41–55.

Weyl, H. 1949. *Philosophy of Mathematics and Natural Science*. Princeton, NJ: Princeton University Press.

White, C. 2015. "Establishing Personal Identity in Reincarnation: Minds and Bodies Reconsidered." *Journal of Cognition and Culture* 15: 403–430.

———. 2016. "The Cognitive Foundations of Reincarnation." *Method and Theory in the Study of Religion* 28: 264–286.

Whiten, A., and E. Flynn. 2010. "The Transmission and Evolution of Experimental Microcultures in Groups of Young Children." *Developmental Psychology* 46: 1694.

Whiten, A., J. Goodall, W. C. McGrew, T. Nishida, V. Reynolds, Y. Sugiyama, C. E. Tutin, R. W. Wrangham, and C. Boesch. 1999. "Cultures in Chimpanzees." *Nature* 399: 682–685.

Whiten, A., V. Horner, and F.B.M. de Waal. 2005. "Conformity to Cultural Norms of Tool Use in Chimpanzees." *Nature* 437: 737–740.

Willett, C. L., and B. M. Rottman. 2021. "The Accuracy of Causal Learning over Long Timeframes: An Ecological Momentary Experiment Approach." *Cognitive Science* 45: e12985.

Williams, A. J., and J. H. Danovitch. 2022. "Is What Mickey Mouse Says Impossible? Informant Reality Status and Children's Beliefs in Extraordinary Events." *Journal of Cognition and Development* 23: 323–339.

Wittgenstein, L. 1953. *Philosophical Investigations*. New York: Macmillan.

Woolley, J. D., C. A. Cornelius, and W. Lacy. 2011. "Developmental Changes in the Use of Supernatural Explanations for Unusual Events." *Journal of Cognition and Culture* 11: 311–337.

Woolley, J. D., and V. Cox. 2007. "Development of Beliefs about Storybook Reality." *Developmental Science* 10: 681–693.

Woolley, J. D., and M. Ghossainy. 2013. "Revisiting the Fantasy-Reality Distinction: Children as Naïve Skeptics." *Child Development* 84: 1496–1510.

Woolley, J. D., and J. Van Reet. 2006. "Effects of Context on Judgments Concerning the Reality Status of Novel Entities." *Child Development* 77: 1778–1793.

Wyman, E., H. Rakoczy, and M. Tomasello. 2009. "Normativity and Context in Young Children's Pretend Play." *Cognitive Development* 24: 146–155.

Wynn, K. 1990. "Children's Understanding of Counting." *Cognition* 36: 155–193.

Wynn, K., P. Bloom, A. Jordan, J. Marshall, and M. Sheskin. 2018. "Not Noble Savages after All: Limits to Early Altruism." *Current Directions in Psychological Science* 27: 3–8.

Young, L., J. A. Camprodon, M. Hauser, A. Pascual-Leone, and R. Saxe. 2010. "Disruption of the Right Temporoparietal Junction with Transcranial Magnetic Stimulation Reduces the Role of Beliefs in Moral Judgments." *Proceedings of the National Academy of Sciences of the United States of America* 107: 6753–6758.

Zelazo, P. D., C. C. Helwig, and A. Lau. 1996. "Intention, Act, and Outcome in Behavioral Prediction and Moral Judgment." *Child Development* 67: 2478–2492.

Zuk, M., and G. R. Kolluru. 1998. "Exploitation of Sexual Signals by Predators and Parasitoids." *Quarterly Review of Biology* 73: 415–438.

Zunshine, L. 2008. "Theory of Mind and Fictions of Embodied Transparency." *Narrative* 16: 65–92.

Acknowledgments

The development of imagination is a collaborative process, and so was the development of this book. An early influence was my literary agent Max Brockman, who helped me think through the book's structure and message. My editor Andrew Kinney provided even more feedback on this front, helping me identify what to include as well as what to cut. My conversations with Andrew left me feeling pleased about where the book was going but also apprehensive about how to get there (in a good way). As the book neared completion, I enlisted the help of Matthew Arredondo to illustrate key findings, methods, and instruments. His images bring many of the book's ideas to life.

I am also indebted to dozens of colleagues with whom I've discussed imagination and its development over the years, including Martha Alibali, Flo Anggoro, Sara Aronowitz, Michael Barlev, Dave Barner, Jennifer Barnes, Igor Bascandziev, Sarah Beck, Jesse Bering, Jessica Black, Peter Blake, Paul Bloom, Elizabeth Bonawitz, Daphna Buchsbaum, Luke Butler, Maureen Callanan, Susan Carey, Nadia Chernyak, Micki Chi, Clark Chinn, Andrei Cimpian, Jenn Clegg, Claire Cook, Kathleen Corriveau, Fiery Cushman, Audun Dahl, Judith Danovitch, Felipe De Brigard, Molly Dillon, Yarrow Dunham, Jan Engelmann, Lisa Fazio, Chaz Firestone, Emily Foster-Hanson, Patricia Ganea, Susan Gelman, Dedre Gentner, Tamsin German, Tobi Gerstenberg, Thalia Goldstein, Micah Goldwater, Tia Gong, Alison Gopnik, Brandon Goulding, Larisa Heiphetz,

Emily Hopkins, Zach Horne, Veronique Izard, Ben Jee, Jenn Jipson, Chuck Kalish, Frank Keil, Deb Kelemen, Melissa Kibbe, Josh Knobe, Melissa Koenig, Barbara Koslowski, Tamar Kushnir, Ashley Landrum, Jon Lane, Brian Leahy, Cristine Legare, Kirsten Lesage, Marjaana Lindeman, Kristi Lockhart, Tania Lombrozo, Tania Luhrmann, Edouard Machery, Jake Mackey, Eric Mandelbaum, Katie McAuliffe, David Menendez, Candice Mills, Henrike Moll, Shaylene Nancekivell, Ara Norenzayan, Shaun Nichols, Melanie Nyhof, Angela Nyhout, Kristina Olson, Ayse Payir, Jonathan Phillips, Marjorie Rhodes, Bekah Richert, Steven Roberts, Samuel Ronfard, Karl Rosengren, Josh Rottman, Mark Sabbagh, Anondah Saide, Barbara Sarnecka, Adena Schachner, Lennart Schalk, Michael Schneider, Laura Schulz, Anna Shusterman, Kristin Shutts, Manvir Singh, Erin Smith, Dave Sobel, Mahesh Srinivasan, Christina Starmans, Jess Sullivan, Marjorie Taylor, Tomer Ullman, Neil Van Leeuwen, Carlo Valdesolo, Caren Walker, Deena Weisberg, Kara Weisman, Henry Wellman, Claire White, Jacqui Woolley, Alice Xu, Andrew Young, and Adrian Zwyssig.

I am particularly indebted to three leading scholars of imagination: Paul Harris, Angel Lillard, and Ori Friedman. Paul Harris has been an inspiration to me since graduate school, as both a researcher and a mentor, and his work is fundamental to our understanding of the relation between imagination and cognitive development. Angel Lillard has done similarly transformative work, and I am indebted to her for reading my entire manuscript and providing insightful feedback, particularly with respect to the sections on fantasy and play. Ori Friedman has unique insights on possibility and pretense, and our conversations and collaborations have changed my understanding of both.

I must also acknowledge the James S. McDonnell Foundation, which provided me with the financial support to take time away from teaching to focus on writing, as well as my family, which provided me with the emotional support needed to complete an onerous, multi-year labor of love. My wife, Katie, ensured that, during the pandemic, our house remained sane and functional—a place where we could continue to work as well as play. My son, Teddy, was an ever-ready critic, happy to argue with me about the importance of the research in this book, in particular my own research. And my daughter, Lucy, helped prod me along by calling my book on imagination my "imaginary book," since it didn't yet exist. Now that it does, I look forward to reading it with my family and pointing out the many ways they shaped my thoughts on the functions, and malfunctions, of imagination.

Index